D1371106

The Privatization of Space Exploration

The Privatization of Space Exploration

Business, Technology, Law and Policy

Lewis D. Solomon

Transaction Publishers

New Brunswick (U.S.A.) and London (U.K.)

Copyright © 2008 by Transaction Publishers, New Brunswick, New Jersey.

All rights reserved under International and Pan-American Copyright Conventions. No part of this book may be reproduced or transmitted in any form or by any means, electronic or mechanical, including photocopy, recording, or any information storage and retrieval system, without prior permission in writing from the publisher. All inquiries should be addressed to Transaction Publishers, Rutgers—The State University of New Jersey, 35 Berrue Circle, Piscataway, New Jersey 08854-8042. www.transactionpub.com

This book is printed on acid-free paper that meets the American National Standard for Permanence of Paper for Printed Library Materials.

Library of Congress Catalog Number: 2008009910
ISBN: 978-1-4128-0759-3
Printed in the United States of America

Library of Congress Cataloging-in-Publication Data

Solomon, Lewis D.
  The privatization of space exploration : business, technology, law and policy
  / Lewis D. Solomon.
      p. cm.
  Includes bibliographical references and index.
  ISBN 978-1-4128-0759-3
      1. Space industrialization. I. Title.

HD9711.75.A2S59   2008
338.0919—dc22

                                              2008009910

Contents

1

Introduction

Though millions brush it aside, we actually stand at the historical edge of humanity's serious thrust into outer space. For historians of tomorrow looking back at the twenty-first century, one of [the] most important economic events of all may prove to be the colonization of space and wealth creation beyond our home planet.
—Alvin and Heidi Toffler, Revolutionary Wealth

For decades following Soviet Major Yuri Gagarin's historic flight on April 12, 1961, human spaceflight remained infrequent and the prerogative of the public sector. Prior to June 2004, all space missions involved large, expensive governmental efforts. Then, on June 21, 2004, SpaceShipOne, developed by Elbert (Burt) Rutan's firm Scaled Composites LLC and financed by Microsoft co-founder, Paul G. Allen, became the first privately-funded manned spacecraft to leave the Earth's atmosphere.[1]

In October 2004, the momentum for change occurred, when Rutan-Allen won the $10 million Ansari X Prize for launching two human suborbital flights within two weeks. This achievement jump-started a private sector that went from nowhere to manned spaceflight to the edge of space. Winning the Ansari X Prize, which opened the door to a myriad of possibilities for the private sector space industry, changed the conversation; it got people excited to dream again about human spaceflight. Wealthy entrepreneurs began putting their faith and money into space endeavors. By pioneering in an affordable way through the atmosphere, the winners of the Ansari X Prize pierced the membrane that previously kept private enterprise anchored to the Earth.

Generations from now, if humans live in colonies in space, mine resources on the Moon and asteroids, and tap the sun's power from space, they likely will look back at the SpaceShipOne winning the Ansari X Prize in 2004 as the spark that got them there. It excited those who

1

wanted to dream and it showed that much more is possible than people previously thought.

Space exploration is, however, a risky endeavor, both technologically and financially. When you add legal uncertainty and bureaucratic barriers to the mix, willing investors become far fewer. Nevertheless, a new crop of entrepreneurs has popped up in America, hoping to turn a profit on everything from cheaper cargo launch systems to mining on the Moon and tapping solar energy to space tourism and even celestial hotels. These companies, and undoubtedly other lesser known dreamers, wait anxiously for more defined and nurturing laws as well as streamlined regulations which have been much needed, but long denied to the fledgling commercial space market.

Private exploration and settlement in outer space or on celestial bodies is legal, both in the American and transnational sense. However, the absence of or uncertainty with respect to transnational legal standards under the 1967 Outer Space Treaty[2] regarding property rights for commercial entities, including resource mining and extraction, nevertheless serves as a disincentive to private sector space enterprise. Business entities and investors who are unsure of their rights and lack assurance that their efforts and investments will receive legal protection are understandably hesitant to undertake the risks involved in developing new technologies and investing financial and human resources if they cannot be assured of some reasonable return. Further, uncertainty in the ability to market space resources shrinks the pool of investment capital. Like any other industry, the commercial space market needs a well-defined and minimally regulated legislative regime in order to get off its feet.

Under the 1967 Outer Space Treaty the private sector's efforts and investments may not be legally protected. Now, some forty years later, increased respect exists for market incentives and for the commercialization of previous public sector activities. Nations seeking to encourage private sector space activities ought to revise the Outer Space Treaty (or enter into ancillary treaties) to provide clear transnational legal protection for private property rights and mineral exploitation in outer space. The importance of this one change cannot be over-emphasized. It will not only obviate future disputes but also serve as a prerequisite to outer space development, including solar power generation, lunar mining, and permanently located space habitats as well as lunar and orbital scientific research and manufacturing facilities.

Taking a Step Back and Forward

Space was at the center of America's consciousness in the 1960s. In a memorable September 12, 1962 statement, President John F. Kennedy asserted: "We choose to go to the moon in this decade and do the other things, not because they are easy, but because they are hard."[3] In the spirit of the early 1960s, the United States chose to do the hard, big stuff. Apollo's success in July 1969 made almost anything seem possible, but it did not happen.

The Cold War made spaceflight the province of governmental agencies in the United States. When the Apollo program ended in 1972, the National Aeronautics and Space Administration (NASA) basically became a jobs program and a means to funnel money to a handful of big aerospace firms. NASA could not break away from a launch system based on military technology; it was unable to implement a plan for reliable, safe, low-cost access to space. With high costs have come a slow pace.

Entrepreneurs are now seeking to rescue human space exploration, from NASA's stagnant, decades-long monopoly. By unleashing entrepreneurial activity, it may be possible to change the space paradigm from the historic province of NASA and several giant aerospace firms and lead to a renaissance in human spaceflight. The private sector, not governmental agencies, such as NASA, will likely energize the quest for space exploration and shape the race for the final frontier.

In the first decade of the twenty-first century, private enterprise is bubbling away, threatening to change the space landscape forever. Individuals are designing spacecraft, start-up companies are testing prototypes, and reservations are being taken for suborbital space flights. With for-profit enterprises seeking to carve out a new realm, the exploration and commercialization of space, the development of space as a tourist destination will likely soon become a reality.

The Significance of Space Exploration

The era of space exploration will start when private sector vehicles, developed for this purpose, start taking passengers on suborbital flights, with reusable space travel systems based on low cost, coupled with reliability and safety, but not the highest performance. Suborbital space tourism will provide a starting point and a significant driver for other private sector space endeavors. It offers the potential to create profits, bring in more investors, provide the groundwork and increased funding for research in other space applications, such as low gravity research as well as lunar and asteroid mining, and lead to broader public participation in orbital space travel. Because people would like to spend several days or a week or more in orbit, in addition to economical launch vehicles, the demand for space exploration will drive the construction and operation of space hotels.

The development of passenger space travel turns on the low cost, reliable, safe access to space by means of reusable launch vehicles (RLV). Similar to an airplane, a RLV, as its name indicates, can be used many times. However, a rocket-powered RLV can go into space. Launch cost savings for a RLV seem likely because of three factors. First, reusable passenger launch vehicles will be simpler than the current U.S. Space Shuttle, discussed in Chapter 2, because they will not perform multiple tasks. Second, these crafts will benefit from progress over the past thirty-five years in a number of areas, including materials technology, structural analysis, computer-aided design, and reliability engineering. Third, these vehicles will serve a large market, permitting "airline-like" operations. The increased scale of activity will lower costs and enhance efficiencies. Private firms have incentive to maximize the use of their vehicles by finding more customers, thereby amortizing their nonrecurring costs across an ever expanding consumer base.

The market for space tourism varies, of course, with the price per ticket, with the main obstacles being development of vehicles to fly passengers in a safe, low-cost manner. Surveys show a sufficiently large market in the United States, Canada, Japan, the United Kingdom, and some other Western European nations of those desirous of thrill-seeking recreational experiences to support not only research and development but also business opportunities for reusable launch vehicles.[4] By 2020, estimates indicate that 100,000 individuals a year could be taking suborbital space rides.[5]

Today's space entrepreneurs believe that if you build and operate safe, low-cost vehicles, the market will be there—the customers will come once the transportation systems exist. Large-scale space tourism, beginning with suborbital flights and then visits to the International Space Station or a similar entity would achieve several goals. First, through space exploration middle-class individuals, not multimillionaires, could explore the unknown. Second, it would convince the public that spaceflight is viable—it is safe. Third, it would drive spaceship developers to create even more cost-effective launch vehicles, thereby igniting further corporate interest in space. Fourth, it would create revenue-producing, profit-making businesses, particularly in the field of intellectual property and its spinoffs, which could fund future spacecraft development.

In short, suborbital space exploration would create a new business sector with potential to generate a significant revenue and profit streams, recapture the U.S. public as a large constituent group supporting space programs, and increase the safety and reliability of space vehicles and lower unit costs. By generating the revenues and offering a high volume traffic model needed to justify the capital investment to reduce the cost of space access, the public utilization of space ultimately offers the possibility for large-scale space exploration and utilization as well as the expansion of human settlement into colonies in space.

The Rationale for Space Exploration, Use, and Settlement

Beyond the development of space tourism and the value of exploration for its own sake—humans want to explore and expand into space because they want to—human space exploration serves nine needs—economic and humanitarian.

First, there exists a great potential for significant space-based research projects and technological innovation, including scientific and research laboratories for certain types of physical and chemical phenomena, including microgravity research experiments and agricultural production facilities. With privately-funded launches, companies can keep secret their research and information far easier than with publicly-funded flights.

Outer space provides a near vacuum, little or no gravity, and temperature extremes, offering unique prospects for various chemical and manufacturing processes. Vacuums are useful for industries requiring the crystallization for certain products, such as microchips and pharmaceuticals. The near perfect vacuum of space allows the enhancement of crystal formations, thereby aiding the commercial development of semiconductors and drugs. The near vacuum and weightlessness of space would also aid in manufacturing new and improved pharmaceuticals.

Second, there is the possibility for significant commercial development of the Moon, including tourism, the mining of the lunar surface for minerals and metals, and serving as a laboratory and then a base for long-duration space flights to Mars.[6] With the successful development of low-cost, safe suborbital passenger spacecrafts, lunar tourism, requiring orbital vehicles, could grow as people go to the Moon in ever increasing numbers. Some observers see the growth from suborbital to orbital services taking perhaps ten years, with the step from orbital to lunar travel being even quicker because companies are already studying how to provide lunar tourism services and a strong demand exists based on the attraction of the Moon.[7]

Using the Moon, first as a laboratory and then as a base, rests on readily available lunar resources. Oxygen extracted from lunar rocks and the Moon's polar ice could be used for life support and as a propellant. Hydrogen taken from lunar polar ice could be used as fuel and in various industrial processes. Dealing with a lunar environment different from the Earth's, for example, its low gravity and the exposure to solar and cosmic radiation, seems surmountable.[8]

Apparently, minerals are in abundance on the Moon,[9] either directly on its surface or located only a few feet below. For example, iron is relatively abundant on the Moon and could be separated to a high level of concentration relatively easily. Other lunar materials could be separated into their component elements, such as aluminum, titanium, and magnesium, through various chemical techniques, using some processes similar to those used on Earth and others consisting of novel adaptions. Lunar silicon could be used to construct large solar cell arrays. The isotope helium-3, available on the Moon's surface, is a potential fuel for use in nuclear fusion reactors on Earth or in space. Because of the inefficiencies and the costs associated with launching large amounts of earthly raw materials into space, lunar mining, excavation, and processing operations could generate materials for the construction of large spacecrafts and space stations either on the Moon or from space platforms.

At present, it is difficult to say which, if any, of these possibilities will be commercially and financially viable. We will need to know not only what is available but also how to capture and process what is present.

Beyond the Moon, asteroids beckon with the promise of resources.[10] Many asteroids are rich in water and in water-bearing minerals as well as other minerals. These resources could be extracted and used both for life support and as materials for various types of space infrastructure. A mass driver, a catapult, could hurl mined materials into orbit for construction purposes. Some valuable, lightweight materials found in asteroids, such as platinum, could be shipped back to Earth.

Third, the benefits of space exploration include tapping space solar power to achieve pollution-free power generation, thereby lessening America's (and more generally, developed nations') dependence on expensive, finite, and often-polluting energy sources. Space solar power involves placing large solar energy collectors in orbit with the energy generated used by other space vehicles or transmitted back to Earth. It is possible to harvest solar energy to provide power for space-based structures, such as space platforms, processing plants, and manufacturing facilities.[11]

Beaming energy back to Earth would begin with space-based solar power stations—solar power satellites—collecting the sun's energy where it is available nearly full-time in the geostationary orbit, some 22,300 miles above the Equator. The solar-cell arrays, with several kilometers of lightweight solar cells exposed to sunlight, could convert solar energy to electricity, then convert the electricity to microwave beams directed to a collector array in a secure area on the Earth. The collector could then convert the waves back to electricity. Although solar energy would be available 24/7, the downside involves the energy loss inherent in the transmission to Earth as well as cost considerations. To undersell electricity produced by coal or nuclear power plants, solar power satellites would need to be built from lunar materials, with nearly all of the materials for these satellites likely obtainable from Moon-based materials, such as silicon. Lunar polar ice could be used to make the propellants for spacecraft to transport the materials needed to build the orbiting satellites.

Fourth, supply depots and orbiting platforms could stock enormous quantities of supplies too heavy to be carried on longer-distance explorations. These way stations could also be used for staging massive construction projects.

Fifth, as a result of necessity, emergencies could trigger space exploration; otherwise, humans could be stuck on Earth without any options. In June 2006, the renowned British astrophysicist Stephen Hawking warned: "It is important for the human race to spread out into space for the survival of the species. Life on Earth is at the ever-increasing risk of being wiped out by a disaster, such as sudden global warming, nuclear war, a genetically engineered virus or other dangers we have not yet thought of."[12]

One possibility not mentioned by Hawking is the risk posed to life on Earth by asteroids.[13] Presenting perhaps the greatest natural threat to human civilization, an asteroid, called a meteor once it reaches the Earth's atmosphere, could plow into the Earth and make our planet (or large portions) uninhabitable.

The Earth faces hazardous, possibly catastrophic, asteroid impacts, potential planet killers. In 2029, an asteroid, Apophis, named for an

Egyptian god of destruction, will be visible on Earth to the naked eye. The Earth's gravity may deflect its course, putting it on target for a possible planetary collision in 2036.

By way of disaster prevention, a need exists for deflecting these harmful, even devastating, space rocks. Changing the orbit of asteroids in a controlled manner will enable us to avoid a rare event with huge negative consequences. Spacecraft in orbit over the Earth and/or around the Moon could be equipped with small telescopes capable of searching for earth-crossing asteroids. In addition to these asteroid-watch stations, a gravity tractor, using the gravitation attraction generated by a large, unmanned spacecraft, could, through a controlled deflection, "tow" an asteroid into a safer orbit.[14]

Sixth, the commercialization of space is linked to national security of the United States. Without militarizing space, the U.S. military could leverage private sector advancements in space for defense purposes. For example, reduced launch costs for satellites would help meet the needs of the Department of Defense.

Seventh, the commercialization of space would help the United States maintain its general technological superiority in space, in particular, relative to potential competitors, such as China. A vigorous space exploration program will also help keep, if not reinvigorate, the economic edge currently enjoyed by the United States.

Eighth, new spacecraft enterprises, new spaceports, and related businesses offer jobs for skilled workers. Space activity also generates economic growth and tax receipts.

Ninth, looking far into the future, human space exploration might help solve the mystery of life itself, specifically, why there is life on Earth, but seemingly no place beyond Earth?

Overview of the Book

The dream of human space exploration, use, and settlement remains powerful. Space continues to beckon and mystify. But more is needed than the mere dream of human spaceflight.

Today, entrepreneurial space visionaries, who believe in the dream, plan to make a go of it on their own, not waiting for the United States government to do it, whose efforts are summarized in Chapter 2.

For decades, NASA, a lumbering bureaucracy, begged funds from a reluctant Congress, while continuing its alliance with a handful of big aerospace firms. However, the winds of change are blowing at NASA.

In a key August 2006 move designed to enable the agency to purchase launch services from the private sector sometime after 2010, NASA agreed to support the commercial development of lower cost space vehicles, pioneered by two small companies. Hopefully, in the future NASA will partner with the entrepreneurial private sector, providing the funds through contracts and prizes to help lift human spaceflight, particularly suborbital flights, off the ground and permit it to thrive.

The 2004 Aldridge Commission saw President George W. Bush's vision for space exploration as a sign that the private space industry is about to take off, much like the aviation industry in the early part of the twentieth century. The commercial space market, one that has remained stagnant for decades, needs a shot in the arm, preferably from the young firms with the fresh ideas and enthusiasm that inevitably accompany the idea of turning a profit.

Competition in the marketplace will allow the best concepts with economic potential to emerge. Over the long-term, for successful space exploration, use, and settlement, policymakers and legislators need to take more of the effort away from NASA and turn it over to entrepreneurial for-profit entities, designed to maximize risk-taking and innovation. Once NASA makes (or is forced to make) this commitment, the question that remains is whether the private firms that are showing such promise are ready—technologically, managerially and financially—to compete.

Having a "can do" spirit, today's space entrepreneurs, profiled in Chapters 3 through 6, see private enterprise as a feasible way to achieve a sustained human presence in space. They see life as a spacefaring civilization as a good thing.

In the past, private sector space startups failed because of undercapitalization, a lack of engineering talent, or a reliance on poor technology.[15] Today's space entrepreneurs, innovative and risk-taking, are spending their own funds because they can; many of whom made fortunes in other fields. Firms are sprouting up; what start as billionaires' pet projects soon morph into market contenders for space exploration vehicles and even payload launch services. Already, new technology and cheaper methodologies are being developed by the private firms profiled in this book that are bent on profiting in space.

New business entities, infused with private capital and testing their technology on the open market, are focusing on three areas: regularizing suborbital human space travel through less expensive, but reliable and safe launch vehicles; significantly lowering the cost of launching payloads into space; and creating human habitations in space.

As discussed in Chapter 3, in 2004, SpaceShipOne, a privately-funded spacecraft, rocketed into suborbital space and came safely home. Backed by Paul G. Allen, Burt Rutan's spacecraft won the $10 million Ansari X Prize. Joining with Sir Richard Branson, who founded Virgin Galactic, Rutan is confident he will build the first reliable, commercial spaceliner, a sort of space minivan for space day trips, with scheduled flights before the end of the first decade of the twenty-first century.

As examined in Chapter 4, the Rutan-Branson duo is racing against another American firm, Space Adventures, Ltd., that is backing a Russian spacecraft design and manufacturing company. To date, however, the only self-financed space travelers have been wealthy business executives who paid some $20 million each, through Space Adventures, to fly on Russia's Soyuz missions to the orbiting International Space Station and back. However, now these for-profit enterprises, both Rutan-Branson and Space Adventures, seek to carve out a new realm for space exploration, by promoting the idea of civilian individuals taking trips into the Earth's suborbital realm.

Elon Musk, an Internet tycoon, having co-founded and sold PayPal, runs Space Exploration Technologies Corp. (SpaceX), which he funded, to date, with more than $100 million of his own money. As analyzed in Chapter 5, SpaceX, a fledgling builder of reusable rockets, hopes to reduce the price of launching payloads into orbit. By keeping launch costs down, Musk expects to grow new markets. Even before a successful launch, SpaceX was one of the two firms that received the August 2006 NASA contract worth up to $500 million.

Technology types—Allen and Musk—and space have a connection. Both technology and space are about the future, new discoveries, breaking barriers. Although high tech entrepreneurs push the edge, space has no edge—it is the final frontier. Those who seek to conquer space believe that progress continues unabated.

Chapter 6 considers Robert Bigelow, the near billionaire founder of Budget Suites of America, who, through Bigelow Aerospace, has poured some $75 million into a plan to launch inflatable orbital habitats, aiming at deployment early in the second decade of this century. Bigelow hopes his habitats, once in space, would be expanded and pieced together into rest stops as well as scientific posts and industrial workshops. Recognizing that if he cannot get people and supplies to his habitats his venture is doomed, Bigelow also funded America's Space Prize worth $50 million, hoping that it will act as a catalyst for the development of orbital launch vehicles.

Chapter 7 offers an overview of current transnational and United States laws and regulations dealing with space exploration, use, and settlement. With the prospect of property rights and the ownership of space assets serving as an incentive for the private sector, recommendations are offered for legal changes designed to facilitate private space exploration, use, and settlement. Space will some day be a repository of resources for business as well as a sea of hotels. It is important that transnational law recognize and comport with this eventuality. The chapter concludes with the need to update the 1967 Outer Space Treaty to provide property rights for commercial entities.

Chapter 8 presents a brief conclusion regarding the future of private sector space exploration, focusing on the prospects for the development of a space elevator.

Before considering the significance of the four start-up firms, we turn to the rise, stagnation, and possible revitalization of a federal governmental agency, the National Aeronautics and Space Administration.

Notes

1. John Schwartz, "Manned Private Craft Reaches Space in a Milestone for Flight," *New York Times*, June 22, 2004, A1; William Booth, "Starship Private Enterprise," *Washington Post*, June 22, 2004, A1; *The Economist*, "The Starship Free Enterprise," 371:8381 (June 26, 2004): 79-81.
2. Department of State, United States of America, *United States Treaties and Other International Agreements,* Volume 18, Part 3 (Washington, DC: U.S. Government Printing Office, 1969), 2410-2421, entered into force on October 10, 1967.
3. John F. Kennedy, "Address at Rice University in Houston on the Nation's Space Effort," September 12, 1962, *Public Papers Of The Presidents Of The United States, January 1 to December 31, 1962* (Washington, DC: U.S. Government Printing Office, 1963), 668-671, at 669.
4. P. Collins, "The Coming Commercial Passenger Space Transportation Market," in *The Space Transportation Market: Evolution of Revolution?*, ed. M. Rycroft (Dordrecht, The Netherlands: Kluwer Academic, 2000), 25. See also Patrick Collins, "The Space Tourism Industry In 2030," *Proceedings of Space 2000*, American Society of Civil Engineers, 594-603 <http://www.spacefuture.com/pr/archive/the_space_tourism_industry_in_2030.shtml>(November 1, 2006) and P. Collins, R. Stockmans, M. Maita, "Demand for Tourism in America and Japan, and its Implications for Future Space Activities," 1995, Advances in the Astronautical Sciences, Paper No. AAS 95-605 <http://www. spacefuture.com/pr/archive/demand_for_space_tourism_in_america_ andjapan.shtml>(November 17, 2006).
5. Burt Rutan, "Statement" in *Future Markets For Commercial Space*, Hearing Before The Subcommittee On Space and Aeronautics, Committee On Science, House of Representatives, 109th Congress, 1st Session, Serial No. 109-10, April 20, 2005, 15, 18. One researcher concluded, however, with a suborbital space tourism start date of 2008 and an initial ticket price of $200,000 per trip, passenger numbers could reach 4,000 a year in 2016 and 10,000 by 2020. With 13,000 passengers

in 2021, revenues from suborbital passenger flight are projected at $676 million. Janice Starzyk, "A Fresh Look at Space Tourism Demand," Futron Corp., June 7, 2006. See also Futron Corp., "Suborbital Space Tourism Demand Revisited," August 24, 2006.

6.   See generally Giancarlo Genta and Michael Rycroft, *Space, the Final Frontier?* (New York: Cambridge University Press, 2003), 188-210 and Haym Benaroya, "Prospects of Commercial Activities At a Lunar Base," *Solar System Development Journal* 1:2 (2001): 1-19.

7.   Patrick Collins, "Space tourism: From Earth orbit to the Moon," *Advances in Space Research* 37:1 (2006): 116-122, at 118. See also Patrick Collins, "The Future of Lunar Tourism," November 21, 2003 <http://www.spacefuture.com/archive/the_future_of_lunar_tourism.shtml>(October 23, 2006).

8.   John S. Lewis, *Mining The Sky: Untold Riches From The Asteroids, Comets, And Planets* (Reading, MA: Helix, 1996), 53-56, 58-66, 68-70, 102.

9.   *Ibid.*, 45-50, 73, 102.

10.   *Ibid.*, 88-91, 111-114.

11.   *Ibid.*, 130-140 and Genta and Rycroft, *Space*, 128-136.

12.   Questions and Answers, Professor Stephen Hawking's Press Conference, The Hong Kong University of Science and Technology, June 13, 2006. See also Dennis Overbye, "Stephen Hawking Plans Prelude to the Ride of His Life," *New York Times*, March 1, 2007, A14.

13.   Lewis, *Mining*, 112-113.

14.   Edward T. Lu and Stanley G. Love, "Gravitational tractor for towing asteroids," *Nature* 438:7065 (November 10, 2005): 177-178.

15.   Nathan C. Goldman, *Space Policy: An Introduction* (Ames: Iowa State University Press, 1992), 129-133, summarized the efforts of U.S. private space transportation companies from 1980 to 1990.

2

The Rise, Stagnation, and Possible Revitalization of NASA

From its beginnings in 1958, the National Aeronautics and Space Administration has perpetuated the perception that human space activities, its primary focus, are expensive and risky and thus the federal government and big aerospace must control almost everything to do with space. An iron triangle, consisting of private sector contractors, NASA, and congressional delegations, has developed over the years. Massive programs came to rest on an alignment of technological potential, the federal government's buying power, a space industry hungry for contracts, and political will mainly resting on a quest for and then the preservation of jobs. The federal government and the ever cynical media constantly remind the public how risky and expensive space ventures are, citing the out-of-date Space Shuttle's tragedies and International Space Station's cost overruns. These efforts lead the public to view space ventures as too difficult; therefore, only through the federal government and its astronauts, a few aerospace contractors, and NASA can human spaceflight move forward. However, the report of a 2004 presidential commission and the activity of private sector entities are changing the reality of and possibilities for space exploration.

This chapter surveys the NASA's birth, its glory years, capped by the Apollo 11 lunar landing in July 1969, and the agency's stagnation thereafter. President George W. Bush's vision for returning to the Moon and the 2004 report of the president's Aldridge Commission are then considered.

NASA's Precursor

As the precursor to NASA, the National Advisory Committee for Aeronautics (NACA) was founded in 1915 to close the airplane development technological gap with Europe and, more generally, to promote aviation progress in the United States.[1] The NACA long focused its in-house research and development efforts on innovative airplane technology and military aviation, enabling the domestic private aviation industry to grow successfully. By the mid-1950s, NACA, known for its technical expertise, had some 7,000 employees, modern research facilities cost-ing $300 million, and an annual budget of $100 million.[2] Although a civilian agency, NACA had a close working relationship with the U.S. military, assisting in solving aeronautical research problems and finding applications for its efforts in the civilian sector. However, NACA did not attempt to enter the commercial airline business itself. Despite NACA's aviation-orientation, it eventually branched out into space-related research and development in the 1950s, for example, developing blunt-shaped reentry concept for the recovery of space capsules and the testing the aerodynamics of rocket models.

Preparation in the mid-1950s for the nation's participation in the International Geophysical Year (IGY), an international effort to launch satellites to help collect data on the earth's surface, spurred NACA's space activities. Drawing justification from the upcoming IGY, the United States government began to develop an experimental scientific satellite program, named Project Vanguard, proposed by the National Academy of Sciences and the U.S. Navy. Chosen in 1955, Project Vanguard fore-shadowed American involvement in the future space race with the Soviet Union. As soon as the U.S.S.R. announced that it would launch a satellite into the Earth's orbit, the United States government rushed to get Project Vanguard off the drawing board and onto the launch pad. All of a sudden, the modest committee that helped the military solve aeronautical research problems, which it could translate into civilian applications, found itself countering Soviet space spectaculars, which were not long in coming, and preserving national prestige.

On October 4, 1957, the Soviet Union successfully launched a small, beeping satellite, Sputnik 1, into orbit. One month later, the U.S.S.R. launched Sputnik 2, weighing 1100 pounds, much more impressive than Vanguard's three pounds, and carrying a dog as a passenger. With the American public plunged into a crisis mode, the United States rushed the development of Project Vanguard, and set a launch date for December 1957. However, on December 6, 1957, Vanguard made it no more than three feet into the air before it exploded into flames. By all accounts, the U.S.S.R. seemed to be winning the space race, posing a threat to

America's system of democratic capitalism. Finally, on January 31, 1958, the U.S. Army successfully launched the Explorer satellite, joined two months later by Vanguard 1, signaling America's entrance into the space race.[3]

NASA's Birth and Its Early Years

Although the United States successfully orbited two satellites just months after the Soviets, the damage to national prestige had been done. The emerging political consensus, based on a national feeling of vulnerability, dictated the need for a space-focused, civilian research and development organization to win the next victories in the space race, including beating the Soviets to human spaceflight. On July 29, 1958, President Dwight D. Eisenhower signed into law the National Aeronautics and Space Act of 1958,[4] creating a charter for U.S. civilian aeronautical and space activities and requiring the absorption of NACA into a new federal agency, the National Aeronautics and Space Administration (NASA), subsequently established on October 1, 1958.[5] NASA incorporated not only NACA, but also the Army Ballistic Missile Agency and the space components of the Naval Research Laboratory, among other units. NASA soon got to work as the face of the U.S. space program.

NASA's objective was political in nature, namely, a governmental effort to beat the Soviets in space. NASA did not seek to support a commercial space industry as NACA had with aviation; rather the federal government became involved in the development, ownership, and operation of space transportation vehicles. Also, NASA would do the bulk of its spaceflight work through private sector contractors rather than in-house as NACA had done.

NASA set out an ambitious ten-year plan, one designed to catch up with the Soviet Union's apparent lead in the space race. In 1960, NASA presented its long-range space exploration plan to Congress, a plan that included orbital human spaceflight, lunar probes to measure and photograph the Moon's environment, human spaceflight to the Moon, planetary probes to Mars and Venus, and the development of larger launch vehicles to carry heavier payloads, all for the cost of $1 billion to $1.5 billion per year over ten years.[6] Project Mercury, the component of the ten-year plan designed to put a person in space, achieved its goal in 1962.

However, Project Mercury was again too late when it came to the space race. On April 12, 1961, Soviet cosmonaut Yuri Gagarin became the first human in space, flying Vostok 1 on a one-orbit mission around the earth. Although astronaut Alan B. Shepard, Jr. achieved suborbital flight on May 5, 1961, the U.S. lagged behind the U.S.S.R. in space, only achieving orbital flight on February 20, 1962, when John H. Glenn, Jr. piloted the Friendship 7 into orbit around the Earth on a Project Mercury mission. Although the public outcry in the spring of 1961 was not as strident as it had been after the launch of Sputnik, something had to be done to recommit the nation to success in the space race.

On May 25, 1961, President Kennedy went before a joint session of Congress and unveiled a U.S. commitment to land an American on the Moon before the end of the decade, a goal that neither the U.S. nor the U.S.S.R. could achieve immediately. Indicating that the United States faced extraordinary challenges posed by the Cold War and ought to respond in an unparalleled manner, building on its economic and technological strengths, President Kennedy stated:

> [I]f we are to win the battle that is going on around the world between freedom and tyranny, if we are to win the battle for men's minds, the dramatic achievements in space which occurred in recent weeks should have made clear to us all, as did the Sputnik in 1957, the impact of this adventure on the minds of men everywhere who are attempting to make a determination of which road they should take....
>
> Now it is time to take longer strides—time for a great new American enterprise—time for this nation to take a clearly leading role in space achievement, which in many ways may hold a key to our future on earth....
>
> ...I believe that this nation should commit itself to achieving the goal, before this decade is out, of landing a man on the moon and returning him safely to the earth. No single space project in this period will be more impressive to mankind, or more important for the long-range exploration of space; and none will be so difficult or expensive to accomplish.[7]

Correctly assessing the national mood, Kennedy sought this technological goal in Cold War terms. He wanted to prove that the United States could beat the then Soviet Union to the Moon and thus demonstrate the overall superiority of its overall society.

NASA devised Project Gemini as a precursor to the historic Apollo program. At an eventual cost in excess of $5 billion, Project Gemini fueled the expansion of NASA's operations, with the agency becoming a large, well-funded federal bureaucracy, achieving a monopoly on all U.S. space activities, except ballistic missiles and reconnaissance satellites that remained in the Department of Defense.

Project Gemini developed a two-person spacecraft designed to rendezvous and dock with other orbital vehicles, to achieve perfect reentry and landing, and to gain additional knowledge about the effects of weightlessness of crew members during long duration flights. More than anything, however, NASA created Project Gemini, a technological bridge to Apollo, the yet-to-be-developed spacecraft intended to fulfill President Kennedy's mandate to put a man on the Moon by 1969. As Project Gemini progressed, the NASA budget grew rapidly, from $966.7 million in fiscal 1961, to $1.825 billion in 1962, to $3.674 billion in 1963, to $5.1 billion in 1964, and eventually reached $5.25 billion in 1965, some 5 percent of the federal budget.[8] NASA's full-time workers jumped from 10,000 in 1960 to more than 34,000 in 1965. The number of employees working for NASA contractors soared from 36,500 in 1960 to more than 370,000 in 1965.[9] In short, the space program came to represent a key national priority.

As funds flowed freely to NASA, successes followed. It became a matter of national pride for the United States, facing a formidable competitor, the then Soviet Union. In March 1965, Virgil I. (Gus) Grissom and John W. Young flew Gemini 3. Even more successful flights followed, clearly establishing an American lead in the fight for space firsts, especially human spaceflight, including living and maneuvering in space, rendezvousing and docking, and gaining medical data on weightlessness. Additionally, Project Gemini facilitated fifty-two onboard experiments, over half of them technological but also including scientific and medical experiments.[10]

NASA's Glory Years and a Space Spectacular

Then came Project Apollo, the most dramatic and historic American space development to date and the largest nonmilitary technological project ever undertaken by the United States.[11] At a time when NASA had not only ample funding but also ready political and public support and a high degree of autonomy, Project Apollo represented a complicated endeavor.

To fly a person to the Moon, NASA adopted the Lunar Orbit Rendezvous model because it did not require an entire spacecraft to land on the Moon's surface, but only the lunar module.[12] In this model, the three-person spacecraft would split into a lunar module and a command module while in lunar orbit, with the lunar module carrying two astronauts to the Moon and then returning to the command module; thereafter, the entire craft would return to Earth. Because the lunar module would leave behind a portion of its heavy engine before it left the Moon, this model minimized the mass that had to be launched on the return trip.

The eventual Apollo spacecraft consisted of: a command module, where the astronauts sat during most of the mission; a service module, which held the equipment the astronauts needed; and a lunar module, which separated from the rest of the spacecraft and landed on the Moon. In order to launch the Apollo craft, NASA ultimately used the Saturn V, a three-stage rocket designed for unmanned and manned earth orbit and lunar missions.

Although excitement and anticipation ran high for Project Apollo, the launch of Apollo 1 ended in tragedy—the first fatal accident in the history of the American space program—with the death of three astronauts. On January 27, 1967, Apollo 1 moved through preflight tests with astronauts Gus Grissom, Edward H. White II, and Roger B. Chaffee in the command module. Shortly before liftoff, a flash fire broke through the spacecraft's shell, apparently the result of a spark in the capsule's wiring that quickly became deadly because of the craft's oxygen-rich environment. Flames engulfed the capsule, it took five minutes for ground crew to open the spacecraft's hatch, and by that time, the three astronauts had asphyxiated and were immolated. The next day, NASA administrator James Webb summed up shock that gripped the agency and the nation: "We've always known that something like this would happen sooner or later...who would have thought the first tragedy would be on the ground?"[13]

Although the tragedy delayed plans for human spaceflight for eighteen months, America's space program aggressively pursued its lunar goal, with NASA concentrating on the missions of Apollo 4, propelled by the first unmanned launch of the Saturn V rocket, in November 1967, and Apollo 6, the first test of the Saturn V booster, in April 1968. In October 1968, NASA resumed manned Apollo flights with Apollo 7, accomplishing the first human flight around the Moon with Apollo 8 in December 1968, and carrying out tests of the lunar module with Apollo 9 and Apollo 10. Apollo 11 rocketed into space and on July 20, 1969, Neil A. Armstrong fulfilled President Kennedy's mandate and stepped off the lunar module and onto the Moon's surface.

While on the lunar surface, Armstrong, joined by his fellow astronaut Edwin E. (Buzz) Aldrin Jr., planted the American flag, deployed the scientific experiments they brought with them, drilled core samples, and collected rocks to bring back to Earth for scientific study. The two astronauts spent more than two and half hours on the Moon while command module pilot Michael Collins orbited above.

Without a doubt, the Apollo 11 lunar landing represented the crowning glory of the American space program and the fulfillment of President Kennedy's vision. However, NASA had nowhere to go but down after the successful mission. That said, the Apollo program continued to spark national attention after July 1969, with a return trip to the Moon aboard Apollo 12 in November 1969, and the now famous ingenuity of NASA's astronauts and Houston mission control crew when an oxygen tank ruptured aboard Apollo 13 in April 1970.[14]

After Apollo 13, four more missions were flown under the Apollo program, all return trips to the Moon to take different samples and photographs of the lunar terrain and to test a lunar rover vehicle. In 1972, NASA terminated Project Apollo after spending $23.5 billion, embarking on seventeen missions, and putting twelve men on the Moon.[15] Having accomplished one of the greatest feats in national history, NASA needed a new goal and a new presidential mandate. However, articulating a goal as clearly as President Kennedy proved difficult, and getting the necessary funding, during a time defined by the Vietnam War and the public's increasing indifference to the spaceflight, even more so. Years later, in 1987, former NASA Deputy Administrator Hans Mark summed up the agency's problem: "President Kennedy's objective was duly accomplished, but we paid a price: the Apollo program had no logical legacy."[16] As an insular bureaucracy in search of a mission, NASA then sought self-preservation, doing enough to justify its existence.

NASA's Stagnation

After Apollo's success in 1969, the United States was too distracted by the Vietnam War and the nation's social problems to commit substantial resources to the human exploration of the space frontier as it had in the 1960s. The age of manned space exploration seemed to have passed, or at least taken a backseat to important issues on Earth. "We walk safely among the craters of the moon but not in the parks of New York or Chicago or Los Angeles," noted one observer in 1969.[17] Soon after the Apollo 11 mission to the Moon, President Richard M. Nixon agreed that space exploration was not a national priority. In a statement on March 7, 1970, rejecting space projects conceived "as a series of separate leaps, each requiring a massive concentration of energy and will and accomplished on a crash timetable," he proclaimed that the federal government "...must also recognize that many critical problems here on this planet make high priority demands on our attention and our resources."[18] Despite the clear ebb in political and budgetary support for NASA, the agency forged ahead with its first major project to follow Apollo: Skylab.

In May 1973, NASA launched a 100-ton orbital workshop called Skylab, using a Saturn rocket. Despite various mechanical difficulties on its first mission, Skylab provided a habitat for three missions totaling 171 days, and became the site of almost 300 scientific and technical experiments before it fell out of orbit and burned up on its reentry to Earth in 1979.[19] Despite budgetary constrictions, NASA followed Skylab with a number of successful ventures, including: Viking 1, a vehicle, launched in August 1975, that spent almost a year flying to Mars to plant an orbiting craft around that planet; Project Voyager, which sent flyby spacecrafts to observe and record data on Jupiter, Saturn, Uranus, and Neptune in the late 1970s and 1980s; and Landsat, a series of satellites which provided data on vegetation infestations.[20]

After Apollo, and alongside its various projects and programs, NASA committed to the lofty goal of routine access to space at a relatively economical cost in the form of the Space Shuttle, which ultimately became the central component of the U.S. space program in the 1980s, the 1990s, and the first decade of the twenty-first century. The space shuttles that went into service in 1981 remain NASA's only means to transport humans in space.

NASA officials originally sought to design and build a reusable craft that would make human space travel more cost-effective and payload launches less costly, thereby sparking a spaceflight revolution. NASA initially projected the research and development costs for the Space Shuttle program to be $15 billion, later lowered to $10 billion.[21] Because NASA was not privy to that magnitude of funding in the post-Apollo era, to win political approval, the agency modified the vehicle, cutting its cost in half to $5.5 billion.[22] In January 1972, President Nixon approved the plan for research development, and construction of the Space Shuttle, a reusable vehicle designed to take crews and cargo into Earth's orbit, and its slimmed down budget, with NASA promising a whole new generation in spaceflight, focused on cheap, routine access to space.[23]

In its final form, the Space Shuttle is designed to carry some 45,000 tons into a near-Earth (115 to 250 miles) orbit, able to haul scientific and other types of satellites into orbit for various users, and accommodate a flight crew of up to ten persons, with a seven-person crew more common, for a basic seven-day mission.

The first Space Shuttle to launch was Columbia, the world's first reusable spacecraft, which lifted off successfully on April 12, 1981. When it landed like an aircraft two days later, excitement again filled the American public about space travel, this time with the promise that space travel would become routine and eventually accessible to more than a privileged few astronauts.

However, by January 1986, the shuttle had only flown twenty-four times, far fewer than the sixty flights a year promised by NASA at the program's outset.[24] One of the reasons that the shuttle flew so seldom was that it was not fully reusable, or at least not in the sense implied by NASA. Although the basic vehicle could be reused after a spaceflight, NASA had to expend thousands of work hours and millions of dollars retooling many of the shuttle's hardware between flights. After a shuttle touched down, countless components, including its main and strap-on engines, needed to be replaced, retouched, recalibrated, and rebuilt. Thus, the turnaround time for the shuttle turned out to be months rather than several days, thereby rending it unsuitable for commercial use.

If the lag time between missions revealed a vehicle incapable of a regular, sustained flight schedule as did the far higher development and operating costs than originally projected, NASA suffered a major setback with the Challenger disaster in 1986. On January 28, 1986, a leak in one of the two solid rocket boosters detonated the main liquid fuel tank, killing the seven astronauts aboard during its launch.[25] After the accident, Americans seemed to lose faith in NASA, resulting from the agency's efforts to deflect responsibility. Turning to move mundane tasks, NASA increasingly focused on the safety and reliability of the Space Shuttle fleet. Although the Space Shuttle returned to flight in September 1988, NASA also looked to other endeavors, including carrying out innovative space probes, such as the Magellan Venus radar mapper in 1989, the launch of the Hubble Space Telescope in 1990, which continues to generate impressive scientific data about the universe, and various unmanned probes, both successful and unsuccessful, to Mars, beginning in 1993.[26] Despite the valuable scientific information obtained by these unmanned efforts, public fascination with NASA and space exploration ebbed. Failing to establish the routine access to space that NASA had promised after Apollo, exploding spacecraft and multi-billion dollar fumbling came to characterize the decades-long Space Shuttle program.

Even before the 1986 Challenger disaster, President Ronald W. Reagan, seeking to reinvigorate the space program, made a historic announcement in his State of the Union Address on January 25, 1984. Calling for the development of a new space station, he stated: "America has always been greatest when we dared to be great. We can reach for greatness again. We can follow our dreams to distant stars, living and working in space for peaceful, economic and scientific gain. Tonight I am directing NASA to develop a permanently manned space station and to do it within a decade."[27] After Congress committed $150 million to the Space Station Freedom for the 1985 fiscal year, NASA promptly came up with an $8 billion space station design.[28] Japan, Canada, and the European Space Agency then agreed to participate in the building of the space station, basically a hard-to-get-to research laboratory that would be used to study how humans, among other life forms, react to extended stays in orbit, from scratch. The space station was redesigned again and again, each time getting smaller, cheaper, less capable of facilitating the scientific and technological experiments it once promised, and becoming even more controversial.[29] Many U.S. politicians who had once supported the space station became increasingly weary of its decreased viability, but by 1992 the project had created approximately 75,000 jobs in thirty-nine states, mainly in California, Alabama, Texas, and Maryland.[30] As a form of job protection and congressional pork-barreling, political support continued for building the space station. In November 1993, Russia and the United States announced that they would work together to build the space station, relegating other nations to supporting roles.

In 1998, assembly began on the International Space Station (ISS), the successor to national space stations, such as Russia's Mir. Seen as the cornerstone of cooperation between the United States and Russia, space agencies from sixteen countries, including the United States, Russia, Japan, Canada, and the European Space Agency nations, signed up to contribute to the construction of the ISS. Becoming operational in 2004 with the scheduled completion date in 2010, building the ISS will require more than forty assembly and utilization flights. Marked by enormous cost overruns, the most cited cost for the ISS is upwards of $100 billion,[31] although the exact number remains hard to pin down because of the difficulty in figuring out how much is paid by Russia. Accessible by Russian Soyuz spacecrafts and the U.S. space shuttles, the ISS, with an orbiting altitude of some 220 miles above the Earth, is at once the target of much criticism for its cost, its dwindling viability, and its role as a destination for astronauts and as discussed in Chapter 4, for wealthy "space tourists."

For NASA, the ISS serves as an excuse to keep the Space Shuttles flying, justifying the agency's existence and its budget. Together, the operation of the Space Shuttle and the construction and maintenance of the space station currently consume nearly one half of NASA's annual budget,[32] choking the agency's research and development efforts.

Forced to prioritize its scientific and human space missions in a budget-limited environment, NASA focuses on two objectives.[33] First, it seeks to complete assembly of the ISS by 2010, though it will be over budget and underused. Second, until 2010, it clings to keeping the balky, expensive, antiquated space shuttle fleet aloft, which takes U.S. astronauts to and from the space station. The cost of each space shuttle launch currently ranges from $250 million to $500 million, depending on the estimates and the assumptions made.[34] Furthermore, billions of dollars of NASA funds flow each year, not counting actual launches, to over eighteen thousand individuals on the ground to maintain the Space Shuttle program. With the number of crew at the ISS limited to only three people, the space station cannot conduct any meaningful scientific research. In sum, the Space Shuttle and the ISS are of little scientific use and do not provide a pathway to the Moon and Mars.

NASA's Future Prospects

On January 14, 2004, President George W. Bush delineated his concept for space exploration and discovery, harkening back to expanding humanity's frontier in space, one which ultimately revolves around a manned mission to Mars. The plan came in the wake of NASA's latest disaster, the Columbia accident in February 2003, in which a piece of insulating foam punched a hole in the craft's thermal tiles, causing it to burn up, killing the seven astronauts on board during reentry over the southwest United States.

In his "Vision for Space Exploration," the president announced new goals for the United States space program and for NASA as an agency.[35] These included completion of the ISS by 2010, a refocus of ISS research on the long-term impact of space travel on human biology, a manned mission to the Moon by 2020 at the latest (this would be the first time any human has set foot on the lunar surface since 1972) based upon the next generation's spaceship to replace the Space Shuttle, and extended human expeditions leading to a permanent base on the Moon, which would serve as a springboard to a manned mission to Mars.

Overall, Bush's vision had no official price tag. The United States would go to the Moon and then to Mars, slowly, cheaply. His proposal called for adding $12.6 billion to NASA's budget for the fiscal years 2005-2009, although only $1 billion would represent new money, with over $11 billion reallocated from existing NASA programs.[36]

President Bush did not describe in detail his goal of a human mission to Mars, set a date for its completion, or provide any cost estimates. Designation of the Moon-Mars objective will, however, allow NASA to progress beyond the Space Shuttle, but without a sense of urgency advocated by President Kennedy in 1961. In many ways, the Bush 43 proposal echoed that of his father, who proposed a human mission to Mars, using the Moon as a jumping off point, during his presidency in 1989. The elder President Bush asked NASA to send humans to Mars under what he called the Space Exploration Initiative, and believed it could be done by 2019, one-half century after the Apollo 11 landing on the Moon.[37] In the context of public indifference to the president's proposal, Bush 41's plan went nowhere and sank without a trace.

Hans Mark, the former deputy director of NASA, expressed skepticism regarding Bush 43's choice of goals, emphasizing that his proposal would probably not have the public impact of Kennedy's or even Reagan's. As with President George H.W. Bush's plan for a human mission to Mars, Bush 43 "put [it] on a time schedule that was so long it was beyond anybody's political horizon. Any plan whose culmination is more than a decade is probably doomed to failure,"[38] Mark noted, because people cannot get excited about it and thus it lacks solid political constituencies.

As it turned out, during George W. Bush's presidency, building a constituency among the public and politicians for large technology proved difficult. Bush did not mention it in his January 2004 State of the Union address. Unlike the space race in the context of the Cold War, a permanent lunar base and a human mission to Mars failed to serve a national security purpose and thus lacked broad political support. Given the earthly war on terror, human space travel to Mars was not an issue of national significance.

Following up Bush's 2004 concept, in September, 2005, NASA detailed a $104 billion plan to get astronauts to return to the Moon by 2018, and serve as a stepping stone to Mars.[39] No cost estimates were, however, provided for sending astronauts to Mars.

Marking NASA's most concrete step to fulfill President Bush's vision to return astronauts to the Moon, but continuing the bond between the agency and a few giant aerospace contractors for long-term, high-cost projects, in August 2006, NASA awarded a contract potentially worth more than $8 billion to Lockheed Martin Corp. to build a new generation of U.S. manned spacecraft, Orion, a crew exploration vehicle, for human transport missions beyond low-Earth orbit, to replace the Space Shuttle in 2014, with a human lunar landing planned for 2020. Under the contract, Lockheed Martin will receive $3.9 billion through 2013 to design, develop, test, and evaluate the vehicle for initial spaceflights. A second stage, from 2009 through 2019, will provide $4.25 billion for an unspecified number of manned spaceships to go to the ISS and the Moon.[40]

Apart from NASA's budgetary constraints, the long timeline for the development of a crew exploration vehicle, such as Orion, seems problematic. Starting from a lower technological base, it took only five years to develop the Apollo in the 1960s. Moreover, with space shuttle operations scheduled to end in 2010 and the completion of the crew exploration vehicle in 2014, the United States will lack human spaceflight capability for four years, instead relying on Russia to take people into low-Earth orbit. During this four-year period, U.S. taxpayers will pay for human spaceflight bureaucrats and personnel to remain in place without doing anything, except preparatory endeavors.

Marking the first concrete step in its ambitious plan to resume human space exploration, in December 2006 NASA unveiled plans to set up a small and ultimately self-sustaining solar-powered settlement of astronauts on the Moon's south pole around 2020.[41] The proposal envisions initial stays of a week by a four-person crew, followed by longer visits until power and other supplies are in place to make a permanent presence possible by 2024. The lunar settlement would serve as a way station for space travelers, providing not only a haven but also oxygen and hydrogen mined from the Moon to provide water and liquid fuel. NASA did not place a price tag on what will be an extremely expensive venture, with funds redirected from the costly Space Shuttle and U.S. participation in the International Space Station.

Report of the Aldridge Commission

After making his January 2004 speech outlining a space exploration plan for the United States, President Bush signed an Executive Order creating a Presidential Commission "to obtain recommendations concerning implementation of the new vision for space exploration activities of the United States...."[42] The blue ribbon commission, consisting of nine individuals from government, industry, academia, and the military, later became known as the "Aldridge Commission" after its chairman, Edward C. (Pete) Aldridge, Jr., a former Secretary of the U.S. Air Force, and an official at the Department of Defense and various aerospace companies. The president charged the Aldridge Commission to advise him on how to implement his space proposal, and to recommend a plan of action regarding: the science research agenda for the U.S. space program; the technologies, demonstrations, and strategies that could be used for sustainable human and robotic space exploration; the long-term organization options for NASA in its implementation of the space exploration program; the role of private-sector and international participants in implementing U.S. space policy; and managing implementation of the policy, using the available resources.[43]

In June 2004, the Aldridge Commission presented its report, *A Journey to Inspire, Innovate, and Discover*,[44] to President Bush. The report addressed a number of issues, including budgetary constraints, educational initiatives, and public relations recommendations. It detailed two recommendations of interest: first, the need to restructure and transform NASA; and second, building the foundation for the success of the U.S. space program on a robust private sector industry, based on multiple, competing approaches that would not only create jobs but also drive down costs and foster the faster development of new technology.

The Aldridge Commission saw a continued, prominent role in the U.S. space program for the public sector and NASA. The report recommended the establishment of a Space Exploration Steering Council to oversee America's space effort that would continue to involve numerous federal agencies. Reviewing and refining NASA's priorities, the report envisioned the creation of three new NASA units, namely, a technical advisory board to provide independent advice on technology questions and risk mitigation, an independent, expenditure-estimating organization to ensure cost realism and accuracy, and, building on NASA's research expertise, a research and technology organization would sponsor high risk/high payoff technological advancement, while tolerating periodic failures.[45] In order to make the shift to research and development, the Report envisioned the reconfiguration of NASA's existing centers into entrepreneurial-oriented Federally Funded Research and Development Centers (FFRDCs), which would work with the private sector and seek to stimulate economic development, but be barred from competing with commercial firms to manage production programs.[46] The commission noted that the centers, as constituted, place an insufficient priority on new technology, professional growth, and managerial mobility.

The Aldridge Commission criticized NASA for growing too risk averse, and suggested that instead of shying away from new technology after disasters, such as Challenger or Columbia, NASA should explain to the public that risk remains for innovation and progress. The report noted:

> Spaceflight is difficult, hazardous, and confronted by enormous distances, at least in human terms. Despite extensive safety precautions, during its 144 human space missions the United States has lost 17 astronauts. The pursuit of discovery is a risky business, and it will continue to be so for the foreseeable future. Astronauts know this and accept that space flight is not easy or routine; the American people and its leaders also need to understand and accept that space flight is not easy or routine.[47]

The Commission concluded that so long as the government is honest with its constituents about the risks present in space exploration, the American people will accept these risks as a price to pay for discovery.

The heart of the report focused on NASA's relationship with the private sector and need to develop a vast space industry, an entirely new commercial sector. A history without competition and on an inflation-adjusted diminishing budget created a bureaucratic monster in NASA, one which became too risk averse, whose technology was too expensive, and one that too often stifled private initiatives before they had a chance to grow. Unwilling or unable to break its ties with giant aerospace firms, such as Boeing and Lockheed Martin, NASA had no way of driving costs down because it would not allow smaller, private firms to compete for its contracts. As a result, U.S. taxpayers came to foot the bill for larger and larger projects with dwindling success rates. The slowness in developing lower cost, reusable spacecraft may have resulted more from insufficient motivation, not insufficient technology.

The Aldridge Commission report rejected the traditional approach to private industry that had long ruled NASA. Instead of a federal agency running a centralized space program with contract work given to Boeing, Lockheed Martin, and a handful of other aerospace giants, an approach marked by cost overruns and schedule slippages, the commission envisioned a space program dominated by diverse commercial interests, using a trial-and-error marketplace approach, with the business of delivering people and payloads into space handled by the private sector.

In Recommendation 3-1, one of the report's key recommendations, the Aldridge Commission stressed the imperative of private sector involvement:

> The Commission recommends NASA recognize and implement a far larger presence of private industry in space operations with the specific goal of allowing private industry to assume the primary role of providing services to NASA, and most immediately in accessing low-Earth orbit. In NASA decisions, the preferred choice for operational activities must be competitively awarded contracts with private and non-profit organizations and NASA's role must be limited to only those areas where there is irrefutable demonstration that only government can perform the proposed activity.[48]

Furthermore, the Aldridge Commission in Recommendation 5-1, advised, "...NASA aggressively use its contractual authority to reach broadly into the commercial and nonprofit communities to bring the best ideas, technologies, and management tools into the accomplishment of exploration goals."[49]

Specifically, the Commission recommended that NASA establish performance-oriented goals and then allow private sector firms to compete to reach those objectives.[50] This is especially desirable when it comes to low-Earth orbit launch services. Making such a transition would involve releasing NASA technology to the private sector, and waiting for competition to drive prices down and efficiency up. The report's recommendations would mean that huge aerospace firms, such as Boeing and Lockheed Martin, could no longer "name their price" when it came to contracts with NASA. It would also mean that NASA would have to do an about face when it comes to its relationship with the private sector.

To promote commercial activities, the Commission further encouraged NASA to use prizes to spur the achievement of space missions and technological breakthroughs. It supported NASA's establishment of the Centennial Challenge Program, in honor of the one hundredth anniversary of the Wright Brothers' first flight, which provides up to $50 million in any fiscal year for the payment of cash prizes for advancement of space or aeronautical technologies, with no single cash prize in excess of $10 million without the approval of NASA's administrator. With this program as an encouraging first step to promote the development of private space technology, the Commission urged NASA to go even further, and consider establishing more substantial prize programs, such as a $100 million to $1 billion prize for the first private organization to fly humans to the Moon and sustain them for a fixed period before they return home.[51]

In sum, the Aldridge Commission envisioned a wide open playing field for the U.S. space program. It recommended that NASA transition the primary roles in the space program, initially, those involving low-Earth orbit, to the private industry by competitively awarding contracts to private organizations and retaining only the services that are inherently governmental, such as research and development. If implemented fully, the Aldridge Commission predicted that the new U.S. space program—and a restructured NASA—would create jobs in a new private sector, help develop the technology necessary to secure American military strength and economic security, and maintain U.S. long-term competitiveness in the international technology sector.

A New Direction for NASA

Seeking to implement some of the key recommendations of the Aldridge Commission, in November 2005, NASA formed its Commercial Crew/Cargo Project Office to spur private industry to provide cost-effective access to low-Earth Orbit.[52] NASA sought a public-private partnership in which the agency would help pay for orbital transportation capability projects that may lead to the procurement of vehicles to carry cargo and crew to and from the International Space Station after the projected retirement of the Space Shuttle in 2010. NASA became a partner in developing a new commercial transportation vehicle, with the agency as an investor that also is an interested customer.

Subsequently, in August 2006, NASA took the unprecedented step, breaking with the agency's tradition. It awarded contracts under its Commercial Orbital Transportation Services Program to two startup space firms.[53] It signed Space Act Agreements[54] with Space Exploration Technologies Corp., run by commercial space pioneer Elon Musk (Chapter 5), and Rocketplane-Kistler, to develop and demonstrate vehicles, systems, and operations of reliable, low-cost access via spacecraft to a low-Earth orbit. NASA's grant of seed money marked the start of fulfilling President Bush's January 2004 directive to promote private sector participation in space exploration and the National Aeronautics and Space Administration Authorization Act of 2005, which called on the agency to advance space commerce.[55]

Under the agreements, demonstrations are scheduled to begin as early as 2008 and continue through 2010 (or later) for the construction of the space vehicles. It will be carried out in two phases, with Phase 1, designed to demonstrate the safe disposal or return of craft that successfully dock at the International Space Station and deliver cargo. A follow-up option to demonstrate crew transportation is planned. Once demonstrated, NASA plans to purchase commercial transportation services to the ISS, competitively in Phase 2.

The two firms selected will receive approximately $500 million but only if they succeed. Payments will be incremental and based on each one's progress against scheduled milestones.

Unlike a traditional NASA procurement arrangement, where the agency issues detailed requirements and specifications for its flight hardware and takes ownership of any vehicle(s) and accompanying infrastructure a contractor produces, under the COTS program, NASA specified only high level goals and objectives, wherever possible, leaving the two firms to make decisions regarding the design, development, certification, and operation of the systems. It encouraged the two companies to obtain additional financing in the private sector for the project and left each free to market the new space transportation services, when developed, to others. NASA also indicated that it would use this model in its future private sector space exploration programs.

NASA's approach may signal the agency's desire to use smaller commercial firms at least for routine cargo hauling and crew transport to the ISS, with longer-range missions to the Moon handled by NASA and one or more aerospace giants, including Lockheed Martin, Boeing, and Northrup Grumman.

Hoping to tap the innovations produced by small firms, in September 2006 NASA created and funded a nonprofit venture capital firm, Red Planet Capital, to invest federal government money in promising, but underfunded, firms. It marked the first time the federal government started a venture capital fund for civilian purposes. The NASA fund, managed by three experienced venture capital veterans selected by the agency through a competitive process, will receive $75 million in federal funds over five years to invest in emerging rocket and space as well as biomedical technologies that may be useful to NASA.[56]

For too long, NASA's culture remained indifferent, if not hostile, to commercial activity. Its culture was insular, defensive of its interests and that of its big aerospace contractors, ill-equipped to foster entrepreneurial business activities. In the early decades of the twenty-first century, perhaps NASA managers will embrace commercialization and innovation. However, it is unclear whether NASA can successfully meld the public sector and the free market. Bogged down by NASA's bureaucracy and the lack of competition, the "robust space-based industry"[57] sought by the Aldridge Commission remains an open question.

The next four chapters examine the increased private sector involvement in space exploration, especially well-financed, smaller firms trying to mimic America's aviation inventors and pioneers of the past. These firms seek to solve the biggest, current impediment to space exploration, namely, the need to drive down the price of launches and develop reliable, reusable vehicles. A viable space exploration industry will help fuel private sector involvement in space exploration, shrink the federal government's monopoly, and NASA's historic reliance on a handful of aerospace firms.

Notes

1. I have drawn on Roger E. Bilstein, *Orders of Magnitude: A History of the NACA and NASA, 1915-1900*, revised edition (Washington, DC: NASA Office of Management, 1989), 1-12, 15-28, 31-48.
2. *Ibid.*, 43, 49.
3. *Ibid.*, 45, 47.
4. Public Law 85-568. The political maneuvering leading to the passage or the act is summarized in Walter A. McDougall,...*The Heavens and The Earth: A Political History of the Space Age* (New York: Basic Books, 1985), 172-176.
5. Bilstein, *Orders*, 49, 54-59; McDougall, *Heavens*, 195-198; Roger D. Launius, *NASA: A History of the U.S. Civil Space Program* (Malabar, FL: Krieger, 1994), 29-41.
6. NASA Office of Program Planning and Evaluation, "NASA Long Range Plan, 1959," December 16, 1959 <http://www.nq.nasa.gov/ office/pao/History/report59. html> (November 1, 2006). See also Bilstein, *Orders*, 49.
7. John F. Kennedy, "Special Message to the Congress on Urgent National Needs," May 25, 1961, *Public Papers of the Presidents of the United States: John F. Kennedy; January 20 to December 31, 1961* (Washington, DC: U.S. Government Printing Office, 1962), 396-406, at 403, 404. For background on Kennedy's determination to achieve superiority in the Cold War see Launius *NASA*, 56, 58-66; McDougall, *Heavens*, 302-304, 305-306, 315-322; Andrew Chaikin, "Feud at NASA Headquarters Sparked JFK Meeting, Former Deputy Says," Space.com (August 27, 2001) <http://www.space.com/ seamans_jfk_010827-1.html>(February 21, 2007). <http://www.space.com/seamans_jfk_010827-1.html>.
8. Bilstein, *Orders*, 69; Launius, *NASA*, 82; W.D. Kay, *Defining NASA: The Historical Debate over the Agency's Mission* (Albany, NY: State University of New York Press, 2006), 86.
9. Kay, *Defining NASA*, 86; Launius, *NASA*, 70.
10. Bilstein, *Orders*, 77; Launius, *NASA*, 82.
11. I have drawn on Bilstein, *Orders*, 78-81, 88-91, Launius, *NASA*, 84-86.
12. Launius, *NASA*, 76.
13. Bilstein, *Orders*, 80; Launius, *NASA*, 87.
14. Bilstein, *Orders*, 98-99; Launius, *NASA*, 91.
15. Bilstein, *Orders*, 100.
16. Launius, *NASA*, 93.
17. *Ibid.*
18. Richard M. Nixon, "Statement about the Future of the United States Space Program," March 7, 1970, *Public Papers of Richard M. Nixon, 1970* (Washington, DC: U.S. Government Printing Office, 1971): 250-253, at 250, 251.
19. Bilstein, *Orders*, 100-103; Launius, *NASA*, 98-100.
20. Bilstein, *Orders*, 113-114, 129-130; Launius, *NASA*, 101-104.

21.   Launius, *NASA*, 108.
22.   Bilstein, *Orders*, 108; Launius, *NASA*, 108.
23.   Bilstein, *Orders*, 109; Launius, *NASA*, 108-110. For a detailed account of the decision to build the Space Shuttle see T.A. Heppenheimer, *The Space Shuttle Decision: NASA's Search for a Reusable Space Vehicle* (Washington, DC: National Aeronautics and Space Administration, 1999).
24.   Launius, *NASA*, 112.
25.   Bilstein, *Orders*, 134-135; Launius, *NASA*, 115-117.
26.   Bilstein, *Orders*, 139; Launius, *NASA*, 125-129.
27.   Ronald W. Reagan, "Address Before a Joint Session of the Congress on the State of the Union," January 25, 1984, *Public Papers of the Presidents: Ronald Reagan: 1984: Book 1* (Washington, DC: U.S. Government Printing Office, 1986), 87-94, at 90. See also Bilstein, *Orders*, 140; Launius, *NASA*, 119; Howard E. McCurdy, *The Space Station Decision: Incremental Politics and Technological Choice* (Baltimore, MD: Johns Hopkins University, 1990), 53-62, 224-235.
28.   Launius, *NASA*, 119-120.
29.   *Ibid.*, 122.
30.   *Ibid.*, 123.
31.   Greg Klerkx, *Lost In Space: The Fall of NASA and the Dream of a New Space Age* (New York: Pantheon, 2004), 226. See also John Schwartz, "Destination is the Space Station, but Many Experts Ask What for," *New York Times*, December 5, 2006, D1.
32.   Carl E. Behrens, "The International Space Station and the Space Shuttle," CRS Report for Congress, November 14, 2006, Table 1. U.S. Space Station Funding, CRS-5, and CRS-8. See also Klerkx, *Lost*, 140.
33.   For an overview see Behrens, "The International Space Station."
34.   Klerkx, *Lost*, 158.
35.   George W. Bush, "Rematks," 66-68; The White House, "A Renewed Spirit of Discovery: The President's Vision for U.S. Space Exploration," January 14, 2004. See also Eric Pianin, "Bush Outlines Space Agenda," *Washington Post*, January 15, 2004, A1; David E. Sanger and Richard W. Stevenson, "Bush Backs Goal of Flight To Moon To Establish Base," *New York Times*, January 15, 2004, A1; William J. Broad, "Bold Visions, Many Pitfalls," *New York Times*, January 15, 2004, A1; Marcia S. Smith, "Space Exploration: Overview of President Bush's 'Vision for Space Exploration,' and Key Issues for Congress," CRS Report for Congress, December 10, 2004. In August, 2006, President Bush signed a new National Space Policy that supports a Moon-Mars exploration agenda, including the use of space nuclear power systems, and emphasizes ensuring U.S. access to and control of space. The White House, "U.S. National Space Policy," August 31, 2006 (unclassified). See also Marc Kaufman, "Bush Sets Defense as Space Priority," *Washington Post*, October 18, 2006, A1.
36.   George W. Bush, "Remarks at the National Aeronautics and Space Administration," January 14, 2004, *Weekly Compilation of Presidential Documents* 40:3 (January 19, 2004): 68.
37.   George H.W. Bush, "Remarks on the 20th Anniversary of the Apollo II Moon Landing, July 20, 1989, *Public Papers of the Presidents: George Bush: 1989: Book 2* (Washington, DC: U.S. Government Printing Office, 1990), 990-993, at 992. Greg Klerkx traced NASA's prior concepts for human expeditions to Mars in *Lost In Space*, 277-283.
38.   William J. Broad, "Bold Visions, Many Pitfalls," *New York Times*, January 15, 2004, A1.

39. National Aeronautics and Space Administration, Press Release, "NASA Release Plans for Next Generation Spacecraft," Release 05-266, September 19, 2005. See also Warren E. Leary, "NASA Planning Return to Moon Within 13 Years," *New York Times*, September 20, 2005, A1; Guy Gugliotta, "NASA Unveils $104 Billion Plan to Return to the Moon by 2018," *Washington Post*, September 20, 2005, A3; Carolyn Porco, "NASA Goes Deep," *New York Times*, February 20, 2007, A19.

40. National Aeronautics and Space Administration, Press Release, "NASA Selects Orion Crew Exploration Prime Contractor," Release 06-305, August 31, 2006. See also Warren E. Leary and Leslie Wayne, "Lockheed Wins Job of Building Next Spaceship," *New York Times*, September 1, 2006, A1; Renae Merle, "Lockheed Wins Contract to Build NASA's New Spaceship," *Washington Post*, September 1, 2006, A1.

41. National Aeronautics and Space Administration, Press Release, "NASA Unveils Global Exploration Strategy and Lunar Architecture," Release 06-361, December 4, 2006. See also Warren E. Leary, "NASA Plans Permanent Base For Exploration on the Moon," *New York Times*, December 5, 2006, A1; Mark Kaufman, "NASA Plans Lunar Outpost," *Washington Post*, December 5, 2006, A1. For a critique of NASA's plans see Dennis Overbye, "Back to the Moon! But Why?," *New York Times*, December 12, 2006, D1.

42. George W. Bush, Executive Order 13326, "President's Commission on Implementation of United States Space Exploration Policy," January 27, 2004, *Weekly Compilation of Presidential Documents* 40:6 (February 9, 2004): 175-176.

43. *Ibid.*, 175.

44. Report of the President's Commission on Implementation of United States Space Exploration Policy, *A Journey to Inspire, Innovate, and Discover*, June 2004. See also Warren E. Leary, "Panel Says Bush's Space Goals Are Feasible," *New York Times*, June 17, 2004, A21.

45. President's Commission, *Journey*, 7, 24-25.

46. *Ibid.*, 24.

47. *Ibid.*, 15.

48. *Ibid.*, 19. See also *Ibid.*, 7.

49. *Ibid.*, 32.

50. *Ibid.*, 20.

51. *Ibid.*, 32-33. See Recommendation 5-2, *ibid.*, 33. For analysis of prizes, see Molly K. Macauley, "Advantages and disadvantages of prizes in a portfolio of financial incentive for space activities," *Space Policy* 21:2 (May 2005): 121-128; Gregg Maryniak, "When will we see a Golden Age of Spaceflight?," *Space Policy* 21:2 (May 2005): 111-119; David Leonhardt, "You Want Innovation? Offer a Prize," *New York Times*, January 31, 2007, C1. For an overview of NASA's Centennial Challenge program, see NASA, "Introduction to Centennial Challenges," April 21, 2006 <http://exploration.nasa.gov/centennialchallenge//cc_overview_/.litc.html>(December 21, 2006) and Ken Davidson, "Prize Competitions and NASA's Centennial Challenges Program, paper presented at the International Lunar Conference 2005.

52. National Aeronautics and Space Administration, Press Release, "NASA Establishes Commercial Crew/Cargo Project Office," Release 05-356, November 7, 2005.

53. National Aeronautics and Space Administration, "NASA Invests in Private Sector Space Flights with SpaceX, Rocketplane-Kistler," August 18, 2006 and NASA, Press Release, "NASA Selects Crew and Cargo Transportation to Orbit Partners," Release 06-295, August 18, 2006. See also Andy Pasztor, "NASA Selects Two Start-Ups to Build, Operate Spacecraft," *Wall Street Journal*, August 21, 2006,

A2; Dave Ahern, "NASA Selects SpaceX, Rocketplane-Kistler For Cargo Orbiting Demo," *Defense Daily* 231:32 (August 21, 2006): 1.

54.   The National Aeronautics and Space Act of 1958, as amended, specifically 42 USC § 2473(c)(5), permits NASA to enter into a Space Act Agreement, a flexible arrangement, allowing the agency to work cooperatively with private firms to develop and support NASA's mission and U.S. national space priorities.

55.   Public Law 109-155, Section 108.

56.   National Aeronautics and Space Administration, Press Release, "NASA Forms Partnership With Red Planet Capital, Inc.," Release 06-317, September 20, 2006. See also Marc Kaufman, "NASA Invests In Its Future With Venture Capital Firm," *Washington Post*, October 31, 2006, A19.

57.   President's Commission, *Journey*, 47.

3

Building on Winning the Ansari X Prize: Mojave Aerospace Ventures, LLC; Scaled Composites, LLC; Virgin Galactic Airways; and The Spaceship Company

Mojave Aerospace Ventures, LLC (Mojave) funded by billionaire Paul G. Allen and using technology developed by Burt Rutan's firm, Scaled Composites, LLC, achieved spaceflight history, garnering the $10 million Ansari X Prize in October 4, 2004. On September 29, 2004 and October 4, 2004, the latter date, the forty-seventh anniversary of the launch of the then Soviet Union's Sputnik satellite, which kicked-off the U.S.-U.S.S.R. space race, discussed in Chapter 2, Mojave launched a spacecraft, SpaceShipOne, built by Rutan's firm, carrying a pilot and the equivalent weight of two people to a suborbital altitude of 62.5 miles (100 kilometers), twice within two weeks, and returning to Earth.[1] The craft exceeded the altitude requirement on both its scheduled flights as required by the prize competition.

The response to the win was overwhelming. President George W. Bush called from Air Force One, dubbing SpaceShipOne's pilots "true American heros," and celebrating Rutan and Allen for their "spirit of entrepreneurship" and for "opening up the space frontier." George T. Whitesides, the executive director of the National Space Society, a nonprofit education group, reacting to the win, stated, "The right stuff is back. I was just keeping every appendage crossed that I could until [SpaceShipOne] came down." Even Sean O'Keefe, the then top NASA administrator, released a statement saying, "The spirit of determination and innovation demonstrated today show that America is excited about a new century of exploration and discovery."[2]

The Incentive Provided By Prize Money

Prize money, in part, motivated the upstart private sector space endeavor. The $10 million Ansari X Prize, won by Mojave, was modeled on the $25,000 Orteig Prize which inspired Charles Lindbergh's solo, nonstop flight across the Atlantic from New York to Paris in 1927 and launched passenger-carrying commercial aviation. In the early twentieth century, prizes in the aviation and automotive worlds were common. Sometimes prizes were awarded for incremental changes, other times for dramatic breakthroughs, such as Lindbergh's flight.[3] Competing teams often collectively spent far more on research and development than the value of the prize itself.

Dr. Peter H. Diamandis and Gregg E. Maryniak created the X Prize Foundation, a nonprofit educational organization, to utilize competitions as an impetus for progress and innovation. In 1995, the two established the X Prize, which was announced in May 1996, to stimulate the private development of vehicles capable of carrying civilian passengers into space and to end federal government's monopoly on space flight.[4] In addition to promoting the creation of the private sector human space flight industry, they designed the prize to encourage interest in and popularize the commercial use of technology needed for private space exploration and prove that the private sector could outperform a U.S. governmental effort. Given the fascination with low-cost human space travel, some twenty-six teams, worldwide, began spending their own funds designing and building spacecraft and entered the race to win the prize.

Since the age of nine, Peter H. Diamandis, chairman and co-founder of the X Prize, wanted to be an astronaut. As he pursued this goal, earning degrees in aerospace engineering and medicine, he became enamored of Gerald O'Neill's book, *The High Frontier*,[5] and O'Neill's vision of a future in space. However, he realized that NASA ultimately selected only a small fraction of astronaut applications, and only half of the selectees actually flew on the increasingly rare missions to space. Instead of appealing to NASA, Diamandis decided to devote his energy and education to sparking a new private space travel industry. Assuming that the marketplace was real, the technology could be developed, and that suborbital passenger spaceflight represented a stepping stone to the future in space, he and Maryniak, established the X Prize Foundation. He also founded the International Space University in France, a graduate school for the study of space, and co-founded Space Adventures, a firm discussed in Chapter 4. In 2006, he was named the recipient of the first ever $500,000 Heinlein Prize for his contributions to the commercialization of space.

When Rutan-Allen won the prize in 2004, Diamandis was thrilled: "The 9-year-old boy inside me is jumping for joy and waiting to take a flight," he said.[6] He sees a bright future for space tourism because of the willingness of wealthy entrepreneurs, such as Allen, to experiment. "There is sufficient wealth controlled by individuals to start serious space efforts," he said.[7] Furthermore, Diamandis wants to make space travel available to the average citizen: "For the last 30 years, people have thought that space flight is only for a select number of government employees. We want to change that mind-set."[8]

An insurance policy funded the $10 million prize with an insurance company betting against the possibility that any privately-financed venture could succeed. Anousheh Ansari, a former telecommunications entrepreneur and board member of the X Prize Foundation for whom the prize is named, grew up in Iran and came to the U.S. as a teenager. Because becoming an astronaut was not an option for her, she directed her passion for technology first at getting degrees in computer and electrical engineering and then in 1993 at founding a startup telecommunications company, Telecom Technologies, Inc. Ansari and her husband sold their firm for some $750 million to Sonos Networks, in 2000.[9] Possessing considerably more faith in the private sector's ability than the insurance company, in November 2002, Ansari and her brother-in-law, Amir Ansari, paid the policy premium reportedly in excess of two million dollars.[10] Four years later, in September 2006 she became the world's first female, private space adventurer, journeying to the International Space Station, 220 miles above the Earth, on a Russian Soyuz spacecraft.

Mojave's Key Personnel

Mojave combined Rutan's aerospace expertise with Allen's funds to win the $10 million Ansari X Prize. Today, Mojave owns the intellectual property behind winning the prize.

Burt Rutan. Elbert L. (Burt) Rutan, a renowned aircraft designer and space enthusiast, and his firm, Scaled Composites, LLC, based in Mojave, California, have long worked to change the public's mind about private space travel. The technical challenges presented by suborbital flight were nothing new to Rutan.[11]

Rutan, born on June 17, 1943, is the son of a dentist, who was the part owner of an airplane. Rutan and his two older siblings were raised in Dinuba, California (Central Valley), where the boys developed, as his brother Dick recalled, "an abnormal fascination with aviation." While Dick "just wanted to fly," Burt's "particular interest was in the design and structure and the flying products and stuff."[12] An avid model plane builder as a child, at age sixteen Burt soloed a single-engine plane, after logging less than six hours of flight training.

After receiving a B.S. in aeronautical engineering from California Polytechnic State University in San Luis Obispo in 1965, he established an enviable reputation in the world of aeronautics and space engineering. From 1965-1972, Rutan worked as a civilian flight-test project engineer at Edwards Air Force Base in California. After two more years as a test center director for Bede Aircraft Co. in Kansas, in 1974 he returned to California and founded the Rutan Aircraft Factory to develop his own designs for the burgeoning home-built aircraft market. His firm offered customers do-it-yourself aircraft, particularly the popular Vari-Eze, made of lightweight composite materials that would eventually become Rutan's specialty. However, mounting pressures from heightened liability exposure led Rutan to transition out of the home-built aircraft market in the early 1980s.

In 1982, Rutan founded Scaled Composites Inc., to develop proprietary designs and composite scale models of prototype aircraft designs for various commercial and military customers. In 1985, Scaled Composites was purchased by Beech Aircraft Corp., a subsidiary of the Raytheon Co.[13] Then in 1988, Scaled Composites was purchased from Beech Aircraft by a management group headed by Rutan, with financing provided by Wyman-Gordon Co. In turn, Wyman-Gordon entered into a joint venture with Scaled Composites to design and manufacture aerospace composite structures. As part of this deal, Wyman-Gordon became Scaled Composites' principal shareholder.[14] In 2007, Northrop Grumman acquired Scaled Composites.

Prior to winning the Ansari X Prize, Rutan's biggest accomplishment came in 1986, when the Voyager airplane he designed flew nonstop, without refueling, in a record-breaking nine-day flight around the globe. The achievement earned Rutan a Presidential Citizen's Medal, presented by President Reagan on December 29, 1986.

After Voyager's flight, Rutan's firm produced numerous imaginative aircraft designs. He also took on U.S. military contracts, including some research done under maximum secrecy. Throughout the years, he pioneered in the use of lightweight composite materials, almost all for experimental aircraft designs. In 2001, Rutan and ten investors bought back Scaled Composites, returning the firm's ownership to its creator and refocusing its efforts on innovative aircraft designs.[15]

Rutan has long been recognized for his expertise and innovative vision in aeronautics. In June 2003, *Aviation Week and Space Technology* placed Rutan at number 29 on its list of the 100 "stars of aerospace" in the last 100 years,[16] despite the fact that nearly all of his planes were never meant for commercial use. Many of the small firms that planned to market his planes ran into financial or technical problems, preventing Rutan's commercially-designed products from ever getting off the ground. One design, for Beech's Starship, a propeller plan designed to perform like a small jet, fell short of its goals and ultimately proved impractical. Because of its prohibitive price, Beech made only fifty of the planes.[17]

At the start of the twenty-first century, Rutan, financially backed by Paul Allen, rose to the forefront of the private sector space tourism field. On April 1, 2004, the U.S. Federal Aviation Administration's Office of Commercial Space Transportation issued Scaled Composites the agency's first private license for a reusable, suborbital, piloted spacecraft.[18] As a result, Rutan could test-fly SpaceShipOne to the edge of space with the craft becoming the first private, manned vehicle to break the Earth's atmosphere on June 21, 2004. With his company pioneering in licensed private space travel, Rutan noted, "This might even be similar to that wonderful time period between 1908 and 1912 when the world went from a total of 10 airplane pilots to hundreds of airplane types and thousands of pilots in 39 countries." He continued, "We need affordable space travel to inspire our youth."[19]

The 2004 prize-winning flights by SpaceShipOne served as Rutan's latest accomplishment. He pointed to the victory as much-needed progress in commercial space travel. Even before his historic achievement, Rutan commented:

> Thirty years ago, if you had asked NASA—and people did in those days—"How long would it be before I could buy tickets into space?" the answer was, "About 30 years." If you ask today, you'll get about the same answer: 30 years. I think that's unfortunate. There has been no progress at all made toward affordable space travel.[20]

Rutan has elaborated these views on many occasions, including the following statement in June 2004:

> Since Yuri Gagarin and Al Shephard's epic flights in 1961, all space missions have been flown only under large, expensive government efforts. By contrast, [the program developed Scaled Composites and Mojave] involves a few, dedicated individuals who

are focused on making spaceflights affordable. Without the entrepreneur approach, space access would continue to be out of reach for ordinary citizens.[21]

Paul G. Allen. One such entrepreneur is Paul G. Allen, the cofounder of Microsoft Corp. and America's fifth richest person, with an estimated worth of $16 billion as of 2006.[22] Allen funded Rutan's efforts to win the Ansari X Prize in order to fulfill childhood fantasies fueled by science fiction and the U.S. government's space program. "As a child, I read everything I could about space travel," he recalled.[23] Allen, who retains his fascination with technological innovation, acknowledged science fiction played a role in his passion for space, noting, "Science fiction definitely stimulated a large part of my thinking about some of these things."[24] Lauding science fiction as a catalyst for ideas and innovation, he stated: "Science fiction is a big inspiration for creativity and thinking out of the box. It forces you to think about the world and about future possibilities, and it reinforces the idea that creativity can be expressed in many ways through science and technology."[25]

Backing SpaceShipOne reflected Allen's lifelong passions, most importantly, "a love of science and technology, and what can be done with engineering." Recalling the widespread excitement in the 1960s surrounding the NASA space missions—Mercury, Gemini, and Apollo—Allen remembered that he "really got enthralled, and probably more than most kids."[26] When he was eleven, he devoured science fiction novels, including *Rocket Ship Galileo* by Robert A. Heinlein,[27] in which teen entrepreneurs join with a scientist to build and fly a rocket to the moon, and nonfiction books, such as *Rockets, Missiles, and Space Travel* by Willy Ley.[28] Allen, the son of a university library administrator and a teacher, was "completely caught up in [Heinlein's book] and started from there." Even now, Allen remembered devouring science fiction books as an adolescent, noting, "There was (a used bookstore) in the U District [of Seattle]. I would buy used science fiction books for a quarter, or something like that, and read them, and slowly added to my collection that way. I got a pretty broad exposure to the authors of the '60s and '70s."[29]

Allen's first foray into rockets was, however, "inauspicious." When he was twelve years old, he and his cousin tried to build a rocket out of an aluminum armchair leg. Using zinc and sulfur from his chemistry set, Allen got the formula right for the fuel mixture which he packed into the tube, but because he failed to look up the melting point of aluminum, "It made a great noise," Allen recalled, "and then melted into place."[30]

Allen attended Lakeside School with his teenage friend Bill Gates, and went on to Washington State University. In 1975, Allen and Gates created Microsoft Corp., initially a tiny, two-man software partnership in Albuquerque, New Mexico. Even after starting Microsoft, he retained his interest in space. Allen remembered, "At the back of my mind, there was always this desire, this inkling of desire, that someday I would try to do something with aerospace or rocketry."[31] In 1981, Allen took a weekend off from the fledgling software company to attend the launch of the first space shuttle—that of *Columbia*. He recalled that witnessing the launching was worth any tension over the time off with his business partner, fondly remembering the excitement of the launch: "The air basically vibrates. There are hundred of thousands of people yelling, 'Go! Go! Go!'"[32]

Peter Steinbrueck, a Seattle City Council member who has known Allen since elementary school, admires what the billionaire has done for his hometown, despite his public image. "Paul Allen is doing some very good things, but it's tainted by the public image people have of him—this very rich, flamboyant guy who hasn't grown up. He is seen as an enigmatic guy who wants to fulfill his fantasies from childhood," Steinbrueck stated.[33] "When any of us were growing up, there is some kind of underlying dream—to pursue these things when we're older and have a chance to be involved," Allen maintained. He continued, "In terms of sheer coolness few things beat rocketry."[34] Even after he left Microsoft, he retained an interest in science and technology, especially things connected with space and space travel.

Allen retired from Microsoft in 1983, after being treated for Hodgkin's disease, a cancer of the lymphatic system, which has been in remission since 1985, initially retaining more than one quarter of the company's stock, which he has gradually sold. Despite Microsoft's unbelievable success, many of the billionaire's later investments in the "wired world,"[35] melding software, hardware (including telephones), infrastructure (such as cable), and content, were duds. Allen's successful investments included shares of Ticketmaster, (a national computerized ticket service), America Online, and USA Networks, all of which he sold.[36] He still, however, holds a controlling stake in a troubled, debt-ridden cable company, Charter Communications, Inc.[37]

In addition to his post-Microsoft business ventures, Allen is philanthropically-inclined. For example, the Paul G. Allen Family Foundation gave away about $246 million over seven years between 1998 and 2004, including approximately $30 million in 2004. Between the foundation and his personal contributions, Allen had given away more than $800 million through 2004.[38] Believing that technology can cull vast amounts of data to help scientists better understand how the brain works, he contributed $100 million to fund the Allen Institute for Brain Science. The Institute enables neuro-cartographers to go where brain researchers have never gone before, thus far producing the Allen Brain Atlas, a map of a mouse's brain.[39]

In 2000, Allen opened a $240 million interactive museum, the Experience Music Project, originally designed to honor Seattle hero and Allen idol Jimi Hendrix, the 1960s rock and roll legend, but which morphed into a more general celebration of rock music. But Allen's interest in museums is not limited to music. His Experience Music Project shares space with Seattle's Science Fiction Museum and Hall of Fame, which opened in 2004 thanks to Allen's $20 million donation. For Allen, the Science Fiction Museum's exhibits are meant to inspire museum goers and to "plant the seeds of imagination" in others that carried him so far in his own career.[40]

Allen has given back to his hometown in other ways in addition to founding museums. Seattle's major institutions are drawn to Allen's vision of a city built around urban living and biotechnology, as well as his intention to establish a job-creating engine that could rival Boeing Co., which still builds aircraft there. To that end, Allen's Vulcan Development Co. is building a biotechnology hub and housing for over ten thousand people in Seattle.[41]

Fondly remembering the Seattle he once knew growing up—where people walked around downtown—Allen wants a back-to-the-future city, one based upon an authentic, sustainable urban setting. In the sixty acres he owns in the South Lake Union area just outside downtown Seattle, Allen plans to build 10 million square feet of residential and commercial space—condos, apartments, townhouses, a retirement home, biotechnology and medical research facilities, hotels, and retail stores—all connected by trolley car. He imagines a community of scientists never more than a short stroll from their laboratories, in a twenty-first century neighborhood based on high density and green building concepts. Seattle's zoning board approved most of the land use changes Allen sought in the South Lake Union area; public investment in the area is expected to range from $500 million to $1 billion, including infrastructure changes.[42]

An avid fan, Allen has also invested in sports ventures in the Pacific Northwest. He bought the Seattle Seahawks NFL team in 1997 for $200 million.[43] He then invested another $100 million, alongside substantial public investment, toward building the Seahawks' Qwest Field that opened in 2002. Additionally, Allen owns the NBA Portland Trailblazers and the Rose Garden, the complex where the team plays. However, Allen's investments in Portland did not fare well. The Allen-backed firm that owns the Rose Garden and development surrounding it filed for bankruptcy in 2004.

Despite Allen's numerous other projects and investments, he retained a passion for science fiction. For example, when the U.S. government cut funding for the search for extraterrestrial intelligence in 1990s, as the main financial backer, Allen invested $27 million, much of it helping to fund the Allen Telescope Array, which scans other planets and the stars for any signs of civilization. The array, operated by the Search for Extra-Terrestrial Intelligence (SETI) Institute, is located in Northern California and consists of 350 twenty-foot radio dishes, hooked together electrically to form the equivalent of a giant telescope.[44]

Financing The SpaceShipOne Venture

The development of SpaceShipOne and its carrier aircraft, White Knight, cost some $25 million. Rutan defended the budget, noting that his company designed and built the two planes without any help from the space travel establishment. He stated, "[W]e had to develop an entire manned-space program from scratch—our own rocket motor, our own rocket-test facility, and our own flight simulator for training pilots." A team of less than fifty people developed the project, "with absolutely no help from nay-say—excuse me, NASA."[45] Rutan's attitude towards NASA and its extraordinarily expensive shuttle program was infectious after the test flight of the first private manned craft reached space in June 2004. He celebrated with a spectator waving a sign that read "SpaceShipOne. Government Zero."[46]

Through his firm, Vulcan Development Co., Allen funded the SpaceShipOne project. After Rutan turned his attention to designing spacecraft in the late 1990s, in 2001 he quietly began his Tier One program designed to put a manned commercial spacecraft into suborbital flight. At that time, Allen agreed to bankroll the program. Despite the fact that SpaceShipOne had been the frontrunner for the Ansari X Prize for months in 2004, seeing it win impacted its sole financier. "When the rocket fires, your heart jumps right into your throat," said Allen, "It's pure exhilaration."[47]

Rutan and Allen split the Ansari X Prize money, with Rutan distributing part of his share to his team of employees who designed, built, tested, and flew the spacecraft and the launch airplane.[48] Allen sought to recoup the remainder of his funds through licensing arrangements.

The SpaceShipOne Technology

Rutan looked to an airplane-rocket combination, strikingly similar to the method by which the X-15 craft reached suborbital flight forty-one years earlier in July 1963. SpaceShipOne is a lightweight craft built from graphite (a carbon-fiber composite).[49] It is twenty-five feet long, with short, stubby wings and twin vertical tails. The craft makes the first part of its trip attached to the bottom of a jet-powered carrier aircraft, the White Knight. At an altitude of about 50,000 feet, or a little under ten miles, SpaceShipOne separates from the White Knight and glides until the spacecraft's pilot fires its rocket engine and points its nose straight up. Covering the last fifty-three miles to the edge of space takes slightly under two minutes, as the rocket fires for eighty seconds and propels the spaceship in a vertical climb at speeds in excess of 2,500 miles per hour, some four times the speed of sound. After the pilot cuts the motor, the spaceship coasts the remainder of the way to the top of its arc.

Glimpsing the black backdrop of outer space, the thin blue line of the atmosphere, and the curvature of the earth, the pilot encounters weightlessness for more than three minutes as the spacecraft reaches the top of its climb before falling back to Earth. Reconfiguring (more technically "feathering") the craft's wings connected to its twin vertical tails increases the drag as it reenters the atmosphere, acting as a huge air brake and slowing its descent to avoid dangerous overheating. Returning the wings to their original position, the pilot then glides downward and lands the craft similar to an airplane in the same runway where it took off.

SpaceDev, Inc., a publicly-held firm founded by James W. (Jim) Benson, a computer entrepreneur turned spaceman, and Environmental Aeroscience Corp. supplied and tested various components for the Scaled-developed hybrid-engine for SpaceShipOne.[50] The rocket motor burns a combination of hybrid (liquid-solid) fuel consisting of liquified nitrous oxide (laughing gas) and solid synthetic rubber and is the first new rocket engine developed for human space flight since 1972.

During flight, the cabin air is cleaned and pressurized using systems borrowed from hobbyist submarines. Most notably, SpaceShipOne's innovative "feathered" reentry system helps decrease the risk of explosion or fire at the most deadly part of any spaceship's mission. During reentry, SpaceShipOne's wings are folded to provide a "shuttlecock" effect, giving the ship high drag and causing deceleration at a higher altitude.

After the win, Rutan was uncertain about the fate of SpaceShipOne and its partner plane, White Knight. Ultimately, Rutan and Allen donated the record-breaking spacecraft to the Smithsonian National Air and Space Museum, where SpaceShipOne hangs in the Milestones of Flight Gallery next to Lindbergh's Spirit of St. Louis. On winning the Ansari X Prize, Rutan said that he planned to focus all his energy on developing his next model, SpaceShipTwo. "Innovation is what we do here," Rutan noted, "because there's not much else to do in Mojave."[51]

A Licensing Agreement and the Formation of
The Spaceship Company:
A Joint Venture Between Scaled Composites and Virgin Galactic

Rutan turned to designing an improved spacecraft, SpaceShipTwo, for space explorer flights. Passengers will not have to wear oxygen masks or spacesuits because the cabin will be pressurized. They will not only be comfortable, but they also will be provided a spectacular view. There will be sixteen round double-paned and double-sealed windows in the craft for viewing during all stages of flight. These details will make SpaceShipTwo much more marketable to space exploration ventures, such as Sir Richard Branson's Virgin Galactic spaceline. "We've always had a dream of developing a space tourism business and Paul Allen's vision, combined with Burt Rutan's technological brilliance, have brought that dream a step closer to reality." Virgin Galactic's founder Branson, who was knighted in 1999 for promoting entrepreneurship, gushed on September 27, 2004, even before Mojave won the Ansari X Prize. He continued:

> The deals with both [Scaled Composites and Mojave], being announced today, are just the start of what we believe will be a new era in the history of mankind, making the affordable exploration of space by human beings real. We hope to create thousands of astronauts over the next few years and bring alive their dream of seeing the majestic beauty of our planet from above, the stars in all their glory and the amazing sensation of weightlessness. The development will also allow every country in the world to have their own astronauts rather than the privileged few.[52]

Branson, the English entrepreneur who started the privately owned Virgin Group Ltd. in 1970—so named because he had no business experience, with one record store in London—exploded onto the business scene, creating Virgin Music with a recording studio in 1972, Virgin

Atlantic Airways in 1984, and Virgin Mobile in 1999.[53] A conglomerate, the Virgin Group, currently owns stakes in air and rail travel, cell telephones, financial services, music, holidays, soft drinks, and retailing, among other ventures. Virgin Group continues to be a vibrant economic force worldwide, emphasizing branded venture capital and presenting customers with a different entertaining way of doing something, having more than 200 companies and employing approximately 50,000 people in twenty-nine countries.[54] Most importantly, Virgin Group's success—and its capital—has allowed Branson, its multibillionaire founder, to take on innovative investment opportunities, such as space exploration. According to Branson, "[m]y interest in life comes from setting myself huge, apparently unachievable challenges, and trying to rise above them."[55]

In September 2004, Branson formed a new company, Virgin Galactic Airways, the world's first commercial spaceline. Branson, who first registered the Virgin trademark in the area of space travel in 1995 and then registered the Virgin Galactic name in 1999, expects to invest around $100 million in spaceship development and ground infrastructure, and eventually to reinvest the firm's profits in space exploration and the development of a new generation of spacecraft.[56]

At the same he formed Virgin Galactic, Branson's Virgin Group entered into an agreement with Mojave to license the technology from Rutan's Tier One program and develop the world's first privately-funded spaceships dedicated to carrying commercial passengers on suborbital space flights. The September 2004 licensing deal, signed just before SpaceShipOne's first Ansari X Prize flight, focused on eventually sending travelers to suborbital space for relatively affordable fees. The arrangement could be worth up to $22 million to Mojave over fifteen years, depending on how many spacecraft Virgin Galactic orders. In addition to the licensing deal, because Branson agreed to sponsor SpaceShipOne on its historic Ansari X Prize flights, the craft carried the Virgin Galactic logo.[57]

Paul Allen, who made SpaceShipOne's development financially possible, hinted at the vast potential of the September 2004 deal with the Virgin Group, stating: "I backed the development of SpaceShipOne because I saw this as a great opportunity to demonstrate that space exploration could someday be within the reach of ordinary citizens. Today's deal with Virgin represents the next stage in the evolution of the SpaceShipOne concept, and will likely be the first of a number of deals that will utilize the technology developed during its creation."[58]

Subsequently, in July 2005, Virgin Galactic and Scaled Composites formed a joint venture, The Spaceship Company, owned 70 percent by the Virgin Galactic and 30 percent by Scaled Composites, to manufacture spaceships and launch aircraft for suborbital personal spaceflights and market them to spaceline operators, such as Virgin Galactic. On the formation of The Spaceship Company, Branson commented:

> I couldn't be more delighted to announce the formation of this joint venture at the biggest private aviation event in the world. Like many people growing up in the Sixties who witnessed the wonder of man walking on the moon—I dreamt that one day I too would make that "one small step..."! Unfortunately though, over the last three decades, many people gave up hope—luckily people like Burt Rutan never did. His vision has allowed people, like me, to dream again. But even I never dreamed as a boy, that one day, I would form, with Burt, the company which will build the world's first commercial passenger spacecraft![59]

The joint venture's mission is as follows:

> The Spaceship Company plans to make spaceflight affordable for the masses and to demonstrate the commercial viability of manned space exploration. We are dedicated to reaching that goal with the first generation of spaceship systems developed for routine, scheduled flight operations. Those systems will be environmentally friendly and will include new solutions to optimize both safety and the passenger experience. We expect that as the flight hardware matures, and is operated by competing spaceline companies, many thousands of people will experience the wonder of leaving the earth's atmosphere each year.[60]

To implement its mission, The Spaceship Company contracted with Scaled Composites for research, development, testing, and certification of SpaceShipTwo and White Knight Two, with technology licensed from Mojave.

Even with the backing of a seasoned international entrepreneur, the Branson-Rutan joint venture ran into problems regarding U.S. national security. The United States government has implemented export control policies to keep hostile and unstable nations from accessing any American technology that could be used for military purposes. Under the International Traffic in Arms Regulations,[61] enforced by the U.S. Department of State Directorate of Defense Trade Controls (DDTC), the DDTC must clear the exchange of military and dual-use commercial and military information between an entity (or individual) in the U.S. and any entity (or individual) in a foreign nation. This regulatory roadblock initially prevented the U.S.-based Scaled Composites from working with Virgin Galactic, part of the British Virgin Group. But in August 2005, the U.S. State Department removed the barrier by granting Virgin Galactic and Scaled Composites a Technical Assistance Agreement (TAA), allowing the two companies to exchange technical information and fully activate their joint venture.[62] Virgin Galactic president Will Whitehorn, who also serves as Brand Development and Corporate Affairs Director for the Virgin Management Ltd., expressed relief on the TAA's approval, stating, "The thing I find really heartening is the willingness on the part of the U.S. government to make sure that this fledgling industry does prosper in the private sector" and not to "stifle the baby before it's born."[63] Rutan began construction of SpaceShipTwo in April 2006 to specifications created by Virgin Galactic, an active partner in the design process.

SpaceShipTwo will be built by Rutan, based on technologies developed for SpaceShipOne but larger, capable of carrying eight people—two pilots and six passengers—and with more amenities, including large windows for viewing and more comfortable seats to help space tourists withstand the stress of launching. Plans also exist for the development of White Knight Two, a carrier aircraft capable of launching SpaceShipTwo to the edge of space. Although The Spaceship Company's first customer will be Virgin Galactic, it also plans to sell (or lease) the larger spacecraft to other space tourist operations. After the resolution of the U.S. technology transfer issues, Virgin Galactic placed an order to purchase five Space-ShipTwos and two White Knight Twos for $50 million with options on further vehicles, and secured the exclusive use of Scaled Composite's air and spacecraft for the first eighteen months of commercial passenger operations.[64]

Branson is confident that his investments will pay off. "We are filling a gap at NASA. While NASA is equipped to go to Mars and beyond, we're concentrating on taking individuals into space," he noted. Rutan added that the partnership between Virgin Galactic and Scaled Composites will show that space travel is not necessarily "difficult" and most appropriately in the hands of the government.[65] Allen sees private businesses that will carry adventurers briefly into space as a "realistic goal," adding that there are plenty of individuals who would pay for the experience and "you could actually get a return on your investment dollar."[66]

Virgin Galactic anticipates a "three-day experience" for "reasonably fit" space tourists, who seek suborbital space flight. The first day of the space tour will consist of medical tests, a trip briefing, and simulator practice to help get tourist-astronauts accustomed to the conditions aboard SpaceShipTwo. On the next day, passengers will take a practice ride aboard White Knight Two and watch the previous group's launch from a cabin specially designed to look like the interior of SpaceShipTwo. On the third day, the space tourists will get what they came for. During the two and one-half hour flight, they will rocket to the edge of space, experiencing weightlessness and marveling at the view for five minutes before SpaceShipTwo begins its descent.[67]

Early numbers suggest that there are quite a few people who would pay up to $200,000 for such a life changing, albeit brief, experience. Since the founding of Virgin Galactic in 2004, the company received ten of thousands of applications for flight aboard SpaceShipTwo and has taken some two hundred twenty deposits.[68] Branson intends to stick with what works, specifically Virgin Group's international marketing model. To meet what he predicts will be skyrocketing demand, Branson had said he hopes to base each SpaceShipTwo in a different country, including the United States, Australia, Japan, and England.[69] Branson's worldwide approach is only the beginning of his campaign to introduce affordable space tourism to the average citizen.

Despite Branson's plans to scatter the launches of SpaceShipTwos around the world, at present they likely will take off in the United States. New Mexico began construction in 2007 on the groundbreaking $225 million spaceport, Spaceport America, close to the White Sands Missile Range, which will include three intersecting runways and two towers for rocket launches.[70] New Mexico won the chance to provide America's first commercial spaceport by outbidding Florida, Texas, and California, with Virgin Galactic agreeing to pay rent of $1 million a year for twenty years. "I know a little bit about branding," said Branson, "and I think [New Mexico], and [Governor Bill Richardson (D)], will be recognized for a love of adventure, for being willing to take a chance."[71] Branson hopes to complete test flights by the end of 2008 and schedule commercial suborbital trips beginning in 2009 on board a spacecraft dubbed the VSS (Virgin Space Ship) Enterprise.

Rutan and Branson hope to develop and market a spacecraft designed to provide suborbital travel as a safe and affordable vacation option for the middle class. "In the coming era of manned space exploration by the private sector, market forces will spur development and yield new, low-cost space technologies," Rutan asserted. He continued, "If the history of private aviation is any guide, private development efforts will be safer too."[72]

Branson and Rutan's Anglo-America venture competition from Space Adventures, an American firm looking to Russian space technology and expertise. Competition seems inherent in the human genes. It likely will yield the best operational vehicle for suborbital space travel, thereby promoting space tourism based on a demonstrated track record of safety, significantly decreased launch costs to levels needed to meet existing consumer demand, and provide the groundwork to expand the market potential.

Notes

1.   See generally, John Schwartz, "Private Craft Rockets Past Edge of Space," *New York Times*, September 30, 2004, A20; John Schwartz, "Private Rocket Ship Earns $10 Million In New Space Race," *New York Times*, October 5, 2004, A1.
2.   The quotations in this paragraph are from *ibid.*, A1.
3.   Molly K. Macauley, "Advantages and disadvantages of prizes in a portfolio of financial incentives for space activities," *Space Policy* 21:2 (May 2005): 121-128, analyzes the effectiveness of prizes and the role of the government compared with the private sector in administering prizes.
4.   For the history of the Ansari X Prize, see National Aeronautics and Space Administration, History Division, "Ansari X-Prize: A Brief History and Background" <http:history.nasa.gov/ x-prize.htm> (November 20, 2006).
5.   Gerald K. O'Neill, *The High Frontier: Human Colonies in Space*, Third edition (Burlington, Ontario, Canada: Apogee, 2000).
6.   Sandi Doughton, "Houston, We Have a Winner," *Seattle Times*, October 5, 2004, A1. For background on Diamandis, see Eli Kintisch, "Founder's bold dream soars toward reality," *St. Louis Post-Dispatch*, September 22, 2004, A1 and John Schwartz, "Into Space, Without NASA," *New York Times*, August 26, 2003, D1; Paula Berinstein, *Making Space Happen: Private Space Ventures and the Visionaries Behind Them* (Medford, NJ: Plexus, 2002): 122-136.
7.   John Schwartz, "Thrillionaires," *New York Times*, June 14, 2005, D1.
8.   Guy Gugliotta, "A Rocket Flight for the Common Man?," *Washington Post*, June 12, 2004, A1.
9.   For background on Anousheh Ansari see Warren E. Leary, "She Dreamed of the Stars; Now She'll Almost Touch Them," *New York Times*, September 12, 2006, D4; Mark Frachetti, "Star Trek fan is first female space tourist," *Sunday Times* (London), July 23, 2004, 24; Scott Farwell, "Plano family's investment takes off," *Dallas Morning News*, October 10, 2004, 1A; Carol Hymowitz, "Immigrant Couple Use Their Survival Skills To Build Tech Success," *Wall Street Journal*, February 13, 2001, B1. See also Eli Kintisch, "Dreams-turned schemes launch one spaceworthy rocket ship," *St. Louis Post-Dispatch*, September 23, 2004, A1.
10. Scott Farwell, "Plano family's investment takes off," *Dallas Morning News*, October 10, 2004, 1A.
11. For background on Rutan, I have drawn on Bruce V. Bigelow, "Rocket Man," *Christian Science Monitor*, February 11, 2004, 15.
12. *Ibid.*
13. Dow Jones Newswires, Raytheon Unit Acquires Scaled Composites Inc., June 14, 1985. See also *Aviation Week & Space Technology*, "Beech Acquires Scaled Composites, Inc.," 122:25 (June 24, 1985): 19 and *Wall Street Journal*, "Raytheon Unit Acquisition," June 17, 1985, 1.
14. *Aviation Week & Space Technology*, "Beech Aircraft Corp.," 129:20 (November 14, 1988): 46; Press Release, "Wyman-Gordon to establish new manufacturing venture," November 16, 1988; *Aviation Week & Space Technology*, "Rutan's Scaled Composites to Develop Close Support Aircraft, Business Jet," 129:21 (November 21, 1988): 27-28.

15.   *Aviation Week & Space Technology*, "Burt's Buyout," 154:5 (January 29, 2001): 17.

16.   *Aviation Week & Space Technology*, "100 Stars of Aerospace," 158:24 (June 16, 2003): 88-95, at 94. Rutan was *Scientific American's* "Business Leader in Aerospace," 289:6 (December 2003): 63 and one of *Time* Magazine's "100 Most Influential People in the World," 165:16 (April 18, 2005): 103.

17.   Andrew Pollack, "Head in the Clouds," *New York Times*, December 9, 2003, G5 and Bruce V. Bigelow, "Rocket Man," *Christian Science Monitor*, February 11, 2004, 15.

18.   U.S. Department of Transportation, Federal Aviation Administration, Commercial Space Transportation License, LRLS 04-067, Scaled Composites, LLC, April 1, 2004. See also *Houston Chronicle* "Private space travel years away," June 20, 2004, A4.

19.   David Usborne, "US Tests Pave the Way for a New Era of Space Tourism," *The Independent* (London), April 9, 2004, 26.

20.   John Schwartz, "Manned Private Craft Reaches Space in a Milestone for Flight," *New York Times*, June 22, 2004, A1.

21.   Bruce V. Bigelow, "Sky-High Ambition," *San Diego Union-Tribune*, June 14, 2004, A1.

22.   *Forbes*, "The Forbes 400," 178:7 (October 9, 2006): 80-292, at 89.

23.   John Schwartz, "Private Space Travel? Dreamers Hope Catalyst Will Rise From the Mojave Desert," *New York Times*, June 14, 2004, A8.

24.   Andrew Garber, "Allen's dreams fuel new space race," *Seattle Times*, June 20, 2004, A1.

25.   Tomas Alex Tizon, "Where Science, Fiction Meet," *Los Angeles Times*, December 10, 2004, A1.

26.   The quotations are from Schwartz, "Trillionaires." See also Andrew Gumbel, "Countdown to the New Space Race," *The Independent* (London), June 17, 2004, 30.

27.   Robert A. Heinlein, *Rocket Ship Galileo* (New York: Ace, 1947).

28.   Willy Ley, *Rockets, Missiles, and Space Travel* (New York: Viking Press, 1961).

29.   Rebekah Denn, "Paul Allen Boldly Goes Where No Billionaire Has Gone Before," *Seattle Post-Intelligencer*, June 18, 2004, A1.

30.   Schwartz, "Thrillionaires."

31.   *Ibid.*

32.   *Ibid.* See also Laura Rich, *The Accidental Zillionaire: Demystifying Paul Allen* (New York: John Wiley, 2003), 52-53. The inside history of Microsoft is described in Bill Gates, Paul Allen, and Brent Schlender, "Bill & Paul Talk," *Fortune* 132:7 (October 2, 1995): 68-86.

33.   Timothy Egan, "Seahawks Rise, As Does Seattle, In Tycoon's Eye," *New York Times*, January 21, 2006, A1.

34.   John Schwartz, "Private Space Travel? Dreamers Hope Catalyst Will Rise From the Mojave Desert," *New York Times*, June 14, 2004, A8.

35.   Rich, *Zillionaire*, 112. See also David Kirkpatrick, "Over The Horizon With Paul Allen," *Fortune* 130:1 (July 11, 1994): 68-75, at 72.

36.   Rich, *Zillionaire*, 115-117, 121-132, 222.

37.   Roben Farzad, "Charter: Cable's Sucker Stock," *Business Week* 3986 (May 29, 2006): 46-48. See also Rich, *Zillionaire*, 197-204, 206-209, 214-215, 217-221 and George Anders and Peter Grant, "Serious Money," *Wall Street Journal*, January 22, 2004, A1.

38.   Allison Linn, "Microsoft's Co-Founder's Dreams Are Reality," *Seattle Post-Intelligencer*, January 15, 2005, <Lexis-Nexis>.
39.   Nicholas Wade, "Atlas Squeaked: A Complete Map of the Brain of a Mouse," *New York Times*, September 26, 2006, D4. See also Egan, "Seahawks Rise."
40.   Schwartz, "Thrillionaires."
41.   Egan, "Seahawks Rise."
42.   *Ibid.*
43.   Rich, *Zillionaire*, 168.
44.   SETI Institute, "History of SETI," <www.seti.org/site/pp.asp?> (November 13, 2006).
45.   Otis Port, "SpaceShipOne's Heady Flight Path," *Business Week Online*, October 1, 2004. <Lexis-Nexis>.
46.   John Schwartz, "Manned Private Craft Reaches Space in a Milestone for Flight," *New York Times*, June 22, 2004, A1.
47.   Sandi Doughton, "Houston, We Have a Winner," *Seattle Times*, October 5, 2004, A1.
48.   John Schwartz, "Private Rocket Ship Earns $10 Million In New Space Race," *New York Times*, October 5, 2004, A1.
49.   I have drawn on Scaled Composites, "Tier One," <http:www. scaled.com/projects/tierone/info/htm> (May 12, 2005 and November 20, 2006); Guy Gugliotta, "A Rocket Flight for the Common Man?," *Washington Post*, June 12, 2004, A1; Schwartz, "Private Craft Rockets"; John Schwartz, "Private Rocket Ship Earns $10 Million In New Space Race," *New York Times*, October 5, 2004, A1; Paul E. Teague, "Space Cadet," *Design News* 60:4 (March 7, 2005): 82-83. SpaceShipOne flight details are set forth in Michael A. Dornheim, "Reaching 100 Km," *Aviation Week & Space Technology* 161:6 (August 9, 2004): 45-46 and Michael A. Dornheim, "A New Spaceship," *Aviation Week & Space Technology* 160:26 (June 28, 2004): 28-29.
50.   SpaceDev, Inc., U.S. Securities and Exchange Commission, Form 10-KSB, March 28, 2006, 3, 44. See also Bruce V. Bigelow, "$10 Million Space Shot," *San Diego Union-Tribune*, October 5, 2004, A1 and Bruce V. Bigelow, "Deal could help Poway company soar," *San Diego Union-Tribune*, September 20, 2003, C3; Berinstein, *Making Space Happen*, 263-290. Andy Pasztor, "In Race to Take Tourists Into Orbit, Partners Split, Spar," *Wall Street Journal*, February 13, 2007, A1, discussed the feud between Rutan and Benson over who developed the key parts of SpaceShipOne's novel rocket motor. See also email, Burt Rutan to Lewis D. Solomon, June 8, 2007. In September 2006, Benson stepped down as chairman and chief technology officer of SpaceDev to start a commercial space tourism business, Benson Space Co. Andy Pasztor, "New Company Aims to Send Tourists to Space," *Wall Street Journal*, September 28, 2006, B1.
51.   John Schwartz, "Private Rocket Ship Earns $10 Million In New Space Race," *New York Times*, October 5, 2004, A1.
52.   The quotations are from Press Release, "Virgin Group Sign Deal with Paul G. Allen's Mojave Aerospace," September 27, 2004 <http://web.lexis-nexis.com/universe> (May 1, 2006).
53.   Press Release, "Virgin Group Sign Deal." Richard Branson recounts his various business ventures in *Losing My Virginity: How I've Survived, Had Fun, and Made a Fortune Doing Business My Way* (New York: Times Books, 1998), 57-64, 67, 77-83, 89-97, 99-101, 106-109, 116-118, 131-140, 143-167, 179-181, 184-188, 197-212, 219-221, 248-274, 276-281. See also Anthony Failoa, "Flying in the Face of Tradition," *Washington Post*, June 26, 1996, C1; Neil Cavuto, *More Than*

Money: True Stories of People Who Learned Life's Ultimate Lesson (New York: Regan, 2004), 86-99 and Spencer Ross, "Rocket Man," *Wired Magazine* 13.1 (January 2005) <http://wired.com/wired/archive/13.01/branson.html> (October 16, 2006). In September 2006, Branson pledged that his personal profits and other proceeds (such as the sale of shares) from the Virgin Group's airlines and a British rail company, estimated at $3 billion over the next 10 years, would be invested in a host of alternative energy enterprises that do not contribute to global warming. Andrew C. Revkin and Heather Timmons, "Branson Pledges Billions to Help Develop Clean Fuels," *New York Times*, September 22, 2006, C5; Chris Hughes, "Green vision and profits fuel Branson's pledge," *Financial Times* (London), September 23, 2006, 17; Sally Beatty, "Branson's Big Green Investment," *Wall Street Journal*, September 22, 2006, W2.

Then, in February 2007 Branson offers a $25 million prize to anyone who could devise a way to blunt global climate change by removing at least one billion tons of carbon dioxide a year from the Earth's atmosphere. James Kanter, "25 Million to Encourage Clearer Air," *New York Times*, February 10, 2007, B3; Kevin Sullivan, "$25 Million Offered In Climate Challenge," *Washington Post*, February 10, 2007, A13; Eoin O'Carroll, "To cool Earth, just scrub the carbon," *Christian Science Monitor*, February 15, 2007, 12.

54. Press Release, "Branson and Rutan Form 'The Spaceship Company' To Jointly Manufacture and Market Spaceships for the New Sub-Orbital Personal Spaceflight Industry," July 27, 2005 <http://web.lexis-nexis.com/universe> (May 1, 2006).

55. Richard Branson recounts his various business ventures in *Losing My Virginity: How I've Survived, Had Fun, and Made a Fortune Doing Business My Way* (New York: Times Books, 1998), 154-155.

56. Press Release, "Virgin Group Sign Deal." See also Richard Orange, "Hotels in Space," *Knight-Ridder Tribune*, April 2, 2006 <http://web.lexis-nexis.com/universe> (May 9, 2006); Heather Timmons, "Virgin to Offer Space Flights (Even, Sort of, at Discount)," *New York Times*, September 28, 2004, W1; *Wall Street Journal*, "Virgin Group Plans Space Flight," September 28, 2004, B5.

57. Press Release, "Virgin Group Sign Deal"; John Schwartz, "Going to The Moon, Sponsored By M&M's," *New York Times*, October 10, 2004, A12; Timmons, "Virgin To Offer Space Flights."

58. Press Release, "Virgin Group Sign Deal."

59. Press Release, "Branson and Rutan Form 'The Spaceship Company.'" See also Leonard David, "Richard Branson and Burt Rutan Form Spacecraft Building Company," Space.com, July 27, 2005 <http://www.space.com/news/050727_branson_rutan.html> (November 15, 2006).

60. Press Release, "Branson and Rutan Form."

61. 22 Code of Federal Regulations Part 120. See also *Investor's Business Daily*, "Space Race," May 5, 2005, A14.

62. David Leonard, "U.S. State Department Green Lights SpaceShipTwo TAA," *Space News*, August 22, 2005, 6.

63. Leonard, "U.S. State Department."

64. Press Release, "Branson and Rutan Form 'Spaceship Company.'"

65. The Branson and Rutan quotations are from Frances Fiorino, "The Sky's No Limit," *Aviation Week & Space Technology* 163:5 (July 1, 2005): 34. In February 2007, NASA and Virgin Galactic LLC signed a memorandum of understanding to explore a possible collaboration on the development of space suits, hybrid rocket motors, heat shields for spaceships, and hypersonic space vehicles. Memorandum of Understanding Between Virgin Galactic LLC and National Aeronautics

and Space Administration Ames Research Center, February 20, 2007; National Aeronautics and Space Administration, Press Release, "NASA, Virgin Galactic to Explore Future Cooperation," Release: 07-49, February 21, 2007. See also *Space Daily*, "NASA And Virgin Galactic To Explore Future Cooperation," February 22, 2007 <Lexis-Nexis>; *Washington Post*, "Technology," February 22, 2007, D2.

66. John Schwartz, "Thrillionaires," *New York Times*, June 14, 2005, D1.

67. Burt Helm, "Virgin Galactic's Space Odyssey," *Business Week Online*, October 15, 2004 <http://web.lexis-nexis.com/universe> (May 1, 2006).

68. Eleanor Lee, "Somewhere over the rainbow extreme wealth can buy a lifestyle of understated good taste," *Financial Times* (London), September 23, 2006, FT Weekend Magazine, 30.

69. John Schwartz, "Private Rocket Ship Earns $10 Million In New Space Race," *New York Times*, October 5, 2004, A1.

70. T.R. Reid, "New Mexico Plans Launchpad for Space Tourism," *Washington Post*, December 15, 2005; *New York Times*, "British Concern and New Mexico Agree on Building a Spaceport," December 14, 2005, A24; Laura Meckler, "New Mexico Plans First 'Spaceport' For Space Travel," *Wall Street Journal*, December 9, 2005, B1. See also Marc Kaufman, "N.M. County Passes Tax Increase to Fund Spaceport," *Washington Post*, April 6, 2007, A3; Marc Kaufman, "If New Mexico Builds It, Will Space Travelers Come?," *Washington Post*, March 26, 2007, A1; Jeff Kass, "NASA it's not, but desert spaceport nears first launch," *Christian Science Monitor*, September 22, 2006, 1 and John Schwartz, "More Enter Race To Offer Space Tours," *New York Times*, February 18, 2006, B1.

71. T.R. Reid, "New Mexico Plans Launchpad for Space Tourism," *Washington Post*, December 15, 2005.

72. Stephen Pincock, "Fly me to the moon, well, almost." *Financial Times* (London), December 18, 2004, 13.

4

From Space Travel Broker to Joint Venturer: Space Adventures, Ltd.

Unlike Scaled Composites, Space Exploration Technologies, or Bigelow Aerospace, Space Adventures, Ltd., resisted entering the business of manufacturing spacecraft or space habitats. Eric Anderson, the company's president and CEO, explained why: "There's too much risk there."[1] Instead, Space Adventures for much of its history served as a travel broker for the super-rich and space-obsessed. Today, the firm offers two major packages: orbital flights currently beginning at $30 million and suborbital flights for $102,000. It has also entered into joint ventures to market spacecraft and develop spaceports outside the United States. The firm also offers other travel programs.

Orbital Flights

Space Adventures achieved fame by arranging the first civilian flight into space aboard a Russian spacecraft in 2001. The Russians began selling tickets to the International Space Station to raise money for their cash-strapped space program. On April 28, 2001, Dennis A. Tito, an American financier, became the world's first space explorer, at a hefty price tag of $20 million.[2] Following Tito's pioneering flight, to date, four other space explorers, South African Internet entrepreneur, Mark Shuttleworth in 2002, American physicist and businessman, Gregory Olsen in 2005, American telecom entrepreneur, Anousheh Ansari in 2006, and Charles Simonyi, Microsoft software designer in 2007, made trips into orbit to visit the International Space Station aboard the Russian Soyuz craft. They got to wear a spacesuit and spend six to twelve days in orbit.

Space Adventures was all too happy to arrange Tito's 2001 flight aboard the Russian spaceship. "Dennis had the courage, will and adventurous spirit needed to realize his dream of spaceflight," commented

Eric Anderson. "We, at Space Adventures, are proud of our influence in the opening of the space frontier."[3]

Tito, who "dreamed of going to space since [he] was a teenager,"[4] described his flight as "the ultimate exploration experience. It was a privilege to be involved in the first mission of this kind and to lead the way for other private space explorers to fulfill their dream too."[5]

NASA's reaction in 2001 to Tito's desire to go into space reflected its risk-averse nature. Because Space Adventures did not at that time design or build spacecraft, the firm focused on the political roadblocks to space exploration, not technical ones.

After receiving a B.S. in Astronautics and Aeronautics and an M.S. in Engineering Science, Tito worked for NASA's Jet Propulsion Laboratory as an aerospace engineer. He quit after five years, frustrated by the low pay and transitioned from plotting spacecraft trajectories for NASA to life as an investment innovator and manager—the founder and president of Wilshire Associates, an investment management company. Among his other business achievements, he conceived the Wilshire 5,000 Index to track the performance of nearly all equity securities issued by U.S.-headquartered corporations and eventually became wealthy.[6]

With the collapse of the Soviet Union and in need for funds for its space program, Russia offered to sell round trip tickets on a Soyuz spacecraft for a short-duration stay in the International Space Station. For $20 million, plus medical testing and rigorous training, multimillionaires could go to the ISS.

At first, the U.S. government opposed Russia's space tourism venture. The space station was unfinished; it was dangerous and only highly trained astronauts and cosmonauts could go there. The Russians tried to reassure NASA that Tito would receive rigorous screening and training, but the U.S. would not relent. Undaunted, in the summer of 2000, Tito embarked on six months of training.

NASA continued to oppose Tito's trip to the International Space Station. After he signed a contract with the Russians in January 2001, the U.S. agency barred him from entering the Johnson Space Center in March 2001 for training with the cosmonauts and told him that he could not go on the trip, asserting that he had not obtained adequate safety training, his presence would distract the crew and endanger everyone on the craft, and jeopardize space station safety.[7] In addition to NASA's discomfort at being beaten to another space landmark by Russia, the disapproval evidenced the agency's reluctance to endorse any type of private space exploration, forcing the market to base itself in Russia, where revenues from private tourists would help finance that country's space program.

After repeated assurances by the Russians, the United States and the other nations building the ISS relented, but demanded that Tito sign various waivers and enter into liability arrangements. They required him to pledge that he (and his heirs) would not sue NASA if something went wrong. They also required him to agree to pay for anything he broke at the space station.[8]

After the success of Tito's 2001 journey, NASA seemed to become more open to space commercialization and to the idea of space explorers visiting the ISS. In September 2001, NASA issued a draft of its new strategy for the development of space commerce. Although it envisioned opening Space Shuttle flight opportunities to private individuals,[9] NASA continues to limit access to space on its craft to astronauts and cosmonauts.

Space Adventures saw these post-Tito developments as an encouraging sign, noting, "NASA has eased up on the issue and is accepting the inevitability of space tourism."[10] By November 2001, when Shuttleworth announced his intention to visit the space station, the partners in the International Space Station project agreed to his flight and reached an accord as to who would be allowed on the ISS. The agreement covered professional astronauts and cosmonauts as well as spaceflight participants, including tourists.[11]

Daniel S. Goldin, the then NASA administrator, continually warned of the risks of spaceflight as a tourism venture, especially orbital flight. "This is very serious stuff," he said in late 2001. "It is not for the faint of heart. This is not Disneyland."[12] Obsessed with risks, the U.S. Space Shuttle continues to remain off limits to private explorers.

For now, space tourists seeking orbital flight will have to depend on the Russians, who deliver a Soyuz capsule to the International Space Station twice a year, leaving open a seat for a paying tourist or foreign guest researcher. Olsen, the 2005 orbital voyager, reiterated that he was not there to sightsee, but to conduct research in his specialty, infrared optics for low light environmental, astronomical, and military observations. "It's a science mission," he maintained. "I don't expect or anticipate any kind of red carpet treatment. I'll take my orders from the pilot of the ship. I intend to be the best possible scientific space visitor that I can possibly be."[13]

Taking the next step, in July 2006, Space Adventures announced an optional ninety minute spacewalk in a spacesuit for those visiting the International Space Station. Priced at $15 million, the spacewalk will bring the total cost of the trip up to $45 million. To deal with safety concerns, a space walker will be tethered to the spaceship and be accompanied by a cosmonaut.[14]

Looking to the future, by 2011, Space Adventures hopes to offer a trip around the moon in a Russian Soyuz spacecraft for $100 million per customer. The mission, called DSE-Alpha and part of the Deep Space Expeditions (DSE) program, rests on Space Adventures' long-standing partnership with the Russian Federal Space Agency (FSA) and the Rocket and Space Corporation Energia.[15] In 2005, the privatized Russian space agency, Energia, displayed a mock-up of the tiny spacecraft that would be mounted onto a Soyuz rocket and be capable of circumnavigating the moon, as Apollo 8 did in December 1969. The DSE-Alpha mission will begin on a Soyuz craft, which will dock with a booster either in low-Earth orbit or at the International Space Station. The booster will then take the passengers and a Russian pilot the rest of the way around the Moon, including three days navigating around the Moon, and back to Earth. "Space Adventures has the exclusive rights to market and sell the DSE-Alpha mission. We have identified over a thousand people around the world who have the financial resources to participate in an expedition to the moon, but the question remains, who among this group has the sense of exploration and adventure to undertake such a historic mission?," inquired Anderson. He continued, "We have recently spoken with a few of these prospective clients and they are interested and eager to learn more. With this level of interest and enthusiasm, I have no doubt that we'll launch DSE-Alpha by 2010."[16]

The Russian Federal Space Agency expressed happiness to host such an historic mission, especially because of the revenue it will provide for the DSE and other programs. Noting that the required modifications are well-understood and relatively easy to be implemented, Anatoly Perminov, chief of the Russian Federation's FSA, stated, "This commercially funded mission is considered a valuable supplement to the agency's overall goals for a manned spaceflight program."[17] Christopher C. Kraft, a former director of the Johnson Space Center, predicted that the lunar journey will afford amazing views, but less than perfect comfort in a vehicle roughly the size of a sport utility vehicle. "I imagine you could endure that, but man, it would be tough," Kraft noted.[18] Nevertheless,

Anderson is confident that his company will find willing—and financially able—participants for the DSE-Alpha mission, and that it will, in turn, raise space exploration in the international consciousness. "I just love the idea of demonstrating that things can be done for less money than people thought, and paradigms can be shifted," Anderson asserted. "Space flight can be opened up."[19]

Suborbital Flights

Despite Space Adventures' reputation for selling glamourous and even controversial travel packages to the superrich, the company anticipates expanding into the wider commercial market for more affordable suborbital travel. The demand seems to exist. To date, the firm has booked 200 reservations for future suborbital flights with a $10,000 deposit toward the cost of $102,000.[20]

A suborbital trip to be offered by Space Adventures would last sixty to ninety minutes. A pilot would fly passengers sixty-two miles above the Earth, allowing them to experience weightlessness and a fantastic view of the planet. The cost of the trip would include a four-day flight preparation and training program, a much abbreviated version of the nine hundred hours of training that Olsen had to endure in order to visit the International Space Station in 2005. Although this kind of trip does not have the fantastic aura that surrounds orbital flight, it does not carry the current $30 million price tag, making it an ideal package for wealthy Americans with a passion for space travel and a thirst for adventuresome vacations.

Because Space Adventures is essentially a travel broker, not an engineering, design, or manufacturing entity, beginning in 2002, it teamed with another firm to offer suborbital flights. Under the March 2002 arrangement, Russia's Myasishchev Design Bureau, an aerospace manufacturer, would develop a new two-stage reusable passenger spacecraft, the Cosmopolis 21. Similar to Rutan's concept discussed in Chapter 3, the C-21 passenger craft would be affixed to a traditional jet-powered aircraft, the M-55 Geophysika. At nearly thirteen miles aloft, the carrier aircraft would separate so that the passenger module could ignite its rocket engine and propel the module to a height of sixty-two miles.[21]

As the venture has evolved, it will market spacecrafts, called the Explorer Series, designed and manufactured by Myasishchev. The Explorer, a craft derived from Cosmopolis 21, will be launched from a carrier aircraft, the M-55X, much like Rutan's SpaceShipOne craft used the

White Knight airplane. The vehicle will have the capacity to transport up to five people to the edge of space.

In February 2006, Space Adventures and the Federal Space Agency of the Russian Federation agreed that the FSA would oversee and supervise the Explorer's development process. Building on the FSA's leadership in private space travel, Anderson commented, "As they have demonstrated in many past efforts, the Russian space agency's commitment to this new and pioneering project will expedite its eventual success."[22]

Spaceports

Once Mysasishchev perfects the design and manufacture of the Explorer, Space Adventures plans to sell the vehicles to spaceports located around the world. To assure initial bases for its spacecraft and its spaceflight packages, Space Adventures entered into partnerships and joint ventures to build commercial spaceports in various nations, including Singapore and the United Arab Emirates. The proposed $115 million spaceport in Singapore will offer not only a base for suborbital spaceflights, but also astronaut training facilities and parabolic flights to simulate weightlessness. In announcing the spaceport project, Anderson noted, "Singapore is one of the best-connected countries in the world. It is home to one of the world's busiest air and sea ports. Singapore, with its superior geographical and economic infrastructure, is primed to be the hub of a new, revolutionary form of travel—in space."[23] Space Adventures is working with a consortium of Singapore investors and Skeikh Saud Bin Saqr Al Qasimi, the Crown Prince of Ras Al-Khaimah, one of the United Arab Emirates, another of Space Adventures' development partners, to finance the spaceport in Singapore.

Space Adventures also plans develop a spaceport in Ras Al-Khaimah, the United Arab Emirates. Funded by the government of Ras Al-Khaimah—which will put up some $30 million—and Space Adventures itself, among other investors, the spaceport will be located just outside Dubai. "Because of Ras Al-Kaimah's unique airport and spaceport support facilities, His Highness' commitment to space tourism, and the close proximity to Dubai, one of the world's leading luxury tourist destinations, makes it a choice location for spaceflight operations," stated Anderson. "As a global leader of tourism, the United Arab Emirates is an ideal location for a spaceport. Suborbital flights will offer millions of people the opportunity to experience the greatest adventure available, space travel. We are honored to partner with His Highness Sheikh Saud."[24]

Steps to Space

Space Adventures also offers a host of other services, so-called "Steps to Space," all space-related, but not as intensive or expensive as the famous flights to the International Space Station. For example, the firm's Zero Gravity flights, aboard a customized Russian aircraft, Ilyushin-76 MDK, allow passengers to experience the sensation of weightlessness for thirty-second long periods without actually being in space.[25]

Space Adventures: Its Key Personnel and Its Financing

Space Adventures is headquartered in Vienna, Virginia, with an office in Moscow. The company's advisory board includes such space notables as Buzz Aldrin, NASA astronauts Thomas A. (Tom) Jones, Kathy Thornton, Charles Walker, Norm Thagard, Sam Durrance, Byron Lichtenberg, Pierre Thuot, Skylab astronaut Owen Garriott, and Russian cosmonaut Yuri Usachev.[26]

Eric Anderson co-founded Space Adventures in 1998 after interning at NASA in 1995 while a junior in the aerospace engineering program at the University of Virginia.[27] At NASA that summer, Anderson learned that his myopia would prevent him from ever becoming an astronaut. He also observed the choking bureaucracy that would prevent NASA from innovatively putting civilians in space at a reasonable cost. "Things there are massively overinflated," recalled Anderson. "I'd see a million-dollar study produce a 100-page report I could have written in college. The government has no incentive to make things cheaper. The bigger the budget, the more power they wield."[28] Anderson graduated at the top of his UVA class in 1996, and proceeded to raise $250,000 from investors to found Space Adventures. Early investors included Peter Diamandis, cofounder of the X Prize Foundation, and Michael McDowell, who started the Arctic cruise company Quark Expeditions and helped him connect with Russian officials to make Space Adventures a reality. Other investors include: Gloria Bohan, president of Omega World Travel, a local travel agency; Richard Garriott, a gaming software entrepreneur; Randall Rule, a non-aerospace industries venture capitalist; and Dr. Alberto Vasquez, an e-commerce/supply chain management software entrepreneur.[29]

While McDowell helped Anderson hire Russian pilots and lease equipment, he needed no help getting the word out about Space Adventures and its services. At an Explorer Club event in New York, Anderson convinced Lotsie Holton, an heiress to the Anheuser-Busch fortune, and her son and husband to be Space Adventures' first customers on a MIG flight. The

Holtons spread the word about their experience, and Space Adventures soon became popular in wealthy circles as the next outrageous vacation firm for adventure seekers.

In February 2000, Anderson sought out Dennis Tito to ask him to invest in Space Adventures. After they talked at length, Anderson recalled, "Then Tito said, 'I think what you're doing is great—it's the future—but I'm completely uninterested in investing in your company. I do want to orbit the Earth now. Can you help?'"[30] This led to Anderson's negotiations with Russia's underfunded space program and what became Space Adventures' most famous booking: sending an American to the International Space Station aboard a Russian spacecraft all for a mere $20 million.

In addition to the capital its initial and subsequent investors provided the firm, its space travel business has proven profitable. Although a Space Adventures spokeswoman would not disclose the company's profits from serving as a space tour operator, particularly the orbital flights, she assured the curious that "it was enough to keep the lights on."[31]

Evidencing new momentum beginning in 2006, Space Adventures seems poised to go beyond its role as a space travel broker. Whether its joint ventures and partnerships will prove successful, however, remains unclear.

Both Space Adventures and Virgin Galactic want to make space travel more affordable. They offer the near term prospect of competition in suborbital flights in which travelers can experience weightlessness and view the earth's curvature from their craft before returning. By making it a more competitive business it likely will be accessible to more people.

Notes

1.   John Schwartz, "Space Tourists," *New York Times*, October 24, 2004, Section 5, 3.
2.   Dennis A. Tito recounted his adventure in "Expanding the Dream of Human Space Flight," in *Space: The Free-Market Frontier*, ed. Edward L. Hudgins (Washington, DC: CATO Institute 2002), 167-173. See also Patrick E. Tyler, "Space Tourist, Back From 'Paradise,' Lands on Steppes," *New York Times*, May 7, 2006, A3.
3.   Space Adventures, Ltd., Press Release, "Five Years and $120 Million Later, Space Adventures Continues to Drive the Industry," April 28, 2006. See also Space Adventures, Ltd., "Space Adventures Takes First Private Citizen Dennis Tito to the International Space Station," February 6, 2001.
4.   John Schwartz, "For Those Who Can Afford It, 2 New Chances to Fly to Space," *New York Times*, June 18, 2003, A20.
5.   Space Adventures, Press Release, "Five Years and $120 Million Later."
6.   For background on Tito, NASA's intransigency, and his flight, see Greg Klerkx, *Lost In Space: The Fall of NASA and the Dream of a New Space Age* (New York:

Pantheon, 2004), 185-193, 207 and Paula Berinstein, *Making Space Happen: Private Ventures and the Visionaries Behind Them* (Medford, NJ: Plexus, 2002), 402-405. See also Andrew Stuttaford, "Cosmic Capitalist," *National Review Online*, May 1, 2001 <http://www.nationalreview.com/contributors/ stuttafordprint050101. html> (November 15, 2006); Todd S. Purdum, "$20 Million Paid for a Ticket But Tenacity Paved the Way," *New York Times*, April 27, 2001, A1; Warren E. Leary, "Millionaire Hopes to Be First Tourist in Space," *New York Times*, June 20, 2000, A19; Debora Vrana, "Wall St.'s Rocket Scientist," *Los Angeles Times*, June 20, 1999, C1.

7.   *New York Times*, "Tourist Barred, Russia Boycotts NASA Training," March 20, 2001, A21. See also Testimony of Daniel S. Goldin, Administrator, National Aeronautics and Space Administration, *NASA Posture*, Hearing Before The Subcommittee on Space and Aeronautics, Committee on Science, House of Representatives, 107th Congress, 1st Session, Serial No. 107-15, May 2, 2001, 36 and National Aeronautics and Space Administration, Responses to Written Questions Submitted by James T. Walsh, Chairman, *Department of Veterans Affairs And Housing And Urban Development, And Independent Agencies Appropriations For 2002*, Hearings Before The Subcommittee on VA, HUD, and Independent Agencies, Committee on Appropriations, House of Representatives, 107th Congress, 1st Session, Part 1, National Aeronautics and Space Administration, May 3, 2001, 137.

8.   National Aeronautics and Space Administration, Press Release, "International Space Station Partnership Grants Flight Exemption for Dennis Tito," Release 01-83, April 24, 2001. See also Warren E. Leary, "Russia Wins Fight to Be First Space Travel Agent," *New York Times*, April 25, 2001, A4; Warren E. Leary, "Deal Reported in Long-Running Dispute on Putting Tourist on Space Station," *New York Times*, April 21, 2001, A12; Kathy Sawyer, "NASA Grudgingly Clears Millionaire's Spaceflight," *Washington Post*, April 21, 2001, A4.

9.   National Aeronautics and Space Administration, "Enhanced Strategy for the Development of Space Commerce (Draft)," September 24, 2001. See also Leonard David, "Space Tourism—Feasible or Flights of Fancy?," Space.com, October 4, 2001 <http://www.space.com/missionlaunches/space_tourism_01104-2.html> (November 29, 2006).

10.   Mark Carreau, "NASA Approves South African's Vacation in Space," *Houston Chronicle*, December 13, 2001, A8.

11.   International Space Station Multilateral Crew Operations Panel, "Principles Regarding Processes and Criteria for Selection, Assignment, Training and Certification of ISS (Expedition and Visiting) Crewmembers, November 2001, Revision A and European Space Agency, "ISS partners release crew criteria document," February 7, 2002.

12.   Robert Hodierne, "Space Oddity," *Washington Post Magazine*, December 9, 2001, 14-23, at 18.

13.   Mark Carreau, "Scientist Secures His Place in Space," *Houston Chronicle*, March 30, 2004, A1. See also Robert Strauss, "Wait Till You See His Vacation Pictures," *New York Times*, October 23, 2005, 4.

14.   Space Adventures, Ltd., Press Release, "Space Adventures Announces Spacewalk Option for Orbital Spaceflight Clients," July 21, 2006. See also John Schwartz, "Walk in Space for $15 Million (Plus Airfaire)," *New York Times*, July 21, 2006, A12 and Mark Carreau, "Travel Agency Offering Spacewalk for $15 Million," *Houston Chronicle*, July 22, 2006, A17.

15. Space Adventures, Ltd., Press Release, "Space Adventures Offers Private Voyage to the Moon," August 10, 2005. See also *Aerospace Daily & Defense Report*, "Space Adventures to sell Soyuz seats to moon for $100 M each," 215:30 (August 12, 2005): 6; James Bernstein, "Greetings from...The Moon!," *Newsday* (New York), August 11, 2005, A6; John Schwartz, "Private Company Plans $100 Million Tour Around the Moon," *New York Times*, August 10, 2005, A16. For background on Energia see Robert Godwin, *Rocket and Space Corporation Energia: The Legacy of S.P. Korolev* (Burlington Ontario: Apogee Books, 2001).

16. Space Adventures, Press Release, "Space Adventures Offers."

17. *Ibid.*

18. Schwartz, "Private Company Plans."

19. *Ibid.*

20. Kim Hart, "Riches to Ride," *Washington Post*, April 16, 2007, D1. Space Adventures, Ltd. "Company Information, Frequently Asked Questions, Suborbital Spaceflight," <http://www.space adventures.com/company/faq> (November 6, 2006).

21. Space Adventures, Ltd., Press Release, "Sub-Orbital Spacecraft Unveiled in Russia," March 13, 2002. See also Jeremy Watson, "Russians Lead in Space Race with Small Jump for Tourist Kind," *Scotland on Sunday*, March 24, 2002, 24; Mark Carreau, "New High for Adventure Travelers," *Houston Chronicle*, March 15, 2002, A3; Reuters, "Russia Unveils Space Shuttle for Tourists," March 14, 2002.

22. Space Adventures, Press Release, "Space Tourism Pioneers."

23. Space Adventures, Ltd., Press Release, "Space Adventures Announces an Integrated Spaceport Offering Suborbital Spaceflights, Astronaut Training and Interactive Visitor's Center," February 20, 2006. See also Space Adventures, Ltd., Press Release, "Space Adventures To Build Spaceport in Singapore," February 25, 2006; Rob Coppinger, "Singapore, UAE Join Space Race," *Flight International* 169:5025 (February 28-March 6, 2006): 37; Amy Yee, "Spaceport Planned for Singpore," *Financial Times* (London), February 21, 2006, 1; John Schwartz, "More Enter Race to Offer Space Tours," *New York Times*, February 18, 2006, C1.

24. The quotations are from Space Adventures, Ltd., Press Release, "Space Adventures Announces $265 Million Global Spaceport Development Project," February 17, 2006. See also *Space Daily*, "Space Adventures Plans Persian Gulf Spaceport," February 25, 2006 <lexix-nexis>; John Schwartz, "More Enter Race to Offer Space Tours," *New York Times*, February 18, 2006, C1.

25. Space Adventures, "Company Information, Frequently Asked Questions."

26. Space Adventures, Press Release, "Five Years."

27. For background on Anderson I have drawn on James M. Clash, "Space Cowboy," *Forbes* 175:10 (May 9, 2005), 58, 60.

28. *Ibid.*, 60.

29. *Ibid.* and Space Adventures, "Company Information, Current Investors."

30. James M. Clash, "Space Cowboy," *Forbes* 175:10 (May 9, 2005): 60.

31. Jerry W. Jackson, "Company is Planning Spaceflights for Tourists," *Orlando Sentinel*, July 19, 2005, C1.

5

The Quest to Reduce Launch Costs:
Space Exploration Technologies Corp.

Unlike other private space firms, such as Scaled Composites, Elon Musk's Space Exploration Technologies Corp. has not developed a means to put people into space, at least not yet. Instead, SpaceX, as the company is called, has concentrated on breaking into the $4 billion a year global satellite launch game, one largely dominated in the United States by The Boeing Co. and Lockheed Martin Corp., two aerospace giants which receive most of the U.S. government's contracts in this field.

SpaceX seeks to lower the cost of launching satellites and other cargo into orbit by 60 to 70 percent, and then by 90 percent, within ten years, mainly by developing a family of launch vehicles which are lighter, more reusable, and cheaper to build than those in use today. Building on the notion that space launching is overpriced, Musk wants to implement his vision of a low-cost, reliable access to low-Earth orbit for small payloads, whether academic, governmental, or commercial, and help the U.S. private sector, led by his firm, regain its competitive edge in commercial space launches.

After founding SpaceX in 2002, Musk has sunk $100 million of his own money into the firm; whether this investment will pay off for Musk remains uncertain.

SpaceX: Its Origins, Failures, and Successes

With SpaceX, Musk plans to implement the process of "creative destruction," which occurs when new companies enter a market with a cheaper, superior product and force the established firms to improve their product and lower their prices.[1] He seeks to create the first privately-funded startup firm to launch payloads into orbit and then loosen Boeing and Lockheed Martin's hold on the governmental and commercial markets. A 2002 feasibility study convinced Musk that "there is nothing inherently expensive about rockets. It's just that those who have built and operated them in the past have done so with horrendously poor efficiency," resulting from "the bureaucratic tendency to cling to obsolete hardware."[2]

Musk decided to start Space X in June 2002 after analyzing the feasibility study and his discussions with friends that focused on the problems of the stagnant U.S. space business. He began with seven employees, taking top young talent from major aerospace firms in Southern California, who were willing to work long hours to develop the Falcon family of rockets, named after the Millennium Falcon of the "Star Wars" films.[3]

At the present time, SpaceX has some 270 employees with office and manufacturing facilities in El Segundo, California and a 300-acre test site in McGregor, Texas. It has launch complexes at the Vandenberg Air Force Base in California and at the Kwajalein Atoll, in the Marshall Islands, northeast of Australia.[4]

In 2004, Musk indicated that he only had enough funds for three launches. If they all failed, he hinted that would be the end of SpaceX. "If we have three failures in a row, it would be almost impossible to recover," he stated. "If we blow three launches in a row, I don't think anyone will ever want to try another launch with us."[5] By 2006, Musk repeated that three failures in a row is "probably the right number" before he would exit as a space pioneer. However, he added, "But the big driver is customer confidence—will they stick by us? If they will, I'd accept more failures. Customers can count on us to be there in the long run."[6]

A sound business plan motivates Musk and SpaceX employees in their goal of building a low-cost, reliable space launch vehicle. According to X Prize cofounder Peter Diamandis, "The only way you can stimulate a commercial space marketplace is to dramatically lower the costs of getting there. If Musk can provide the same quality of service to space for 25% of the going price, then we will see a shift in the industry."[7] Also, if SpaceX succeeds in sending a rocket into space for as little as one-tenth of the current cost, Musk believes the impact would be huge. "We will absolutely crush the launch business of Boeing, Lockheed and Orbital," he asserted. "Launching rockets is only a small portion of their business, but we will absolutely demolish them."[8]

Bravado aside, Musk has known failures all too well. SpaceX delayed the initial launch of its Falcon 1 rocket by over two years, to November 2005. Subsequently scheduled launches were cancelled several times as a result of various glitches and mishaps.[9] Even then, Musk assured observers that cancelled launches and technical problems were par for the course in developing any new technology. "I don't think the research and development really stops until you've had a few launches or at least a few countdowns so you can refine the process," he noted a few days before one rescheduled, but aborted launch.[10] Then, the March 2006 launch failed. After the planned launch was delayed an hour and a half, the rocket lifted off, only to crash less than one minute into its inaugural flight, resulting in a total loss of the craft. The failure resulted from a cracked aluminum nut that corroded from the sea air during the months of delay allowing fuel to leak, thereby triggering a fire resulting in fuel pressure loss, which shut down the rocket's first-stage engine.[11] Watching the rocket explode, Musk observed "was like being hit in the liver with a blunt object."[12]

In what might have been a ploy to allay fears that the Falcon 1 would never work, Musk remarked that while it would be "ludicrous to call [the failed launch of Falcon 1 in March 2006] a complete success," it would "be just as ludicrous to refer to it as a complete failure." In that vein, Musk pointed to the useful data collected from the failed launch for its sponsor, the U.S. Defense Advanced Research Projects Agency (DARPA). "In that respect the mission was quite successful," he observed.[13]

Some experts brushed aside Falcon 1's failed March 2006 launch attempt. "I think they will ultimately be successful," maintained John R. London III, assistant for project management and development at NASA's Marshall Space Flight Center in Huntsville, Alabama. Despite the launch problems, he continued, "This is not an unusual kind of thing you see, when developing new launch systems."[14]

In a March 2007 test launch, although the Falcon 1 reached an altitude of about 200 miles, it fell short of its intended target altitude and its planned full orbital velocity when its second stage engine shut down early.[15] Musk declared the rocket operational, with another launch scheduled for the second quarter of 2008.

Despite the delays, Musk noted, "No one has even hinted they will even consider leaving SpaceX."[16] Among the hundreds of millions of dollars in advance launch contracts SpaceX garnered,[17] including agreements with NASA, the Pentagon and U.S. Air Force, no one backed out of an order.

SpaceX Contracts

SpaceX has received a number of launch contracts with public sector agencies, including Malaysian Space Agency, the Swedish Space Corporation, and the United States government as well as a few commercial contracts. DARPA awarded SpaceX an $8 million contract under its Falcon Small Launch Vehicle program to develop and demonstrate affordable space launch capability. "With this contract, the Department of Defense is continuing a tradition of supporting new American space launch capabilities," Musk noted. "We are honored to be selected by DARPA and the U.S. Air Force."[18] Under agreement, DARPA funded the first launch of SpaceX's Falcon 1 rocket in March 2006.

In April 2005, SpaceX and Orbital Sciences Corp. jointly received a U.S. Air Force-Space and Missile Systems Center "responsive" multi-launch contract worth up to $100 million to provide low-cost launches of small payloads—presumably military satellites—into space over the next five years, with the work to be completed by April 2010.[19] The term "responsive" means readiness to launch a rocket on twelve months' notice, far shorter than the current standard of several years. The contract was awarded, in part, based on the Falcon 1's projected ability to place a 1,477-pound payload into a 124-mile circular orbit from Cape Canaveral, Florida for some $6 million.

Then, in September 2005, SpaceX signed a contract with a secret party—thought to be a spy agency—for a single, classified launch. Neither SpaceX nor the agency will reveal the parameters of the contract or the type of satellite to be launched.[20]

SpaceX's biggest government contract success, to date, occurred in August 2006. NASA selected SpaceX and Rocketplane-Kistler[21] to split a contract worth approximately $500 million in installments over five years, based on fixed price, performance milestones, to demonstrate their capability to deliver cargo to the International Space Station. As a new approach to help stimulate private funding and potentially reduce government outlays, for the first time, NASA hired contractors to build spacecrafts that private firms, not the U.S. government, will own. Ultimately, NASA will be one of many customers, including for profit businesses and other government agencies, that will purchase the space launches.[22] After NASA's announcement, Peter Diamandis, chairman of the X Prize Foundation, commented, "NASA is the gold standard, and it has given them a stamp of approval."[23] Musk stated, "By stimulating the development of commercial orbital spaceflight, the NASA COTS [Commercial

Orbital Transportation Services] program will have the same positive effect on space travel as the Air Mail Act of 1925 had on the development of safe and affordable air transportation. Moreover, the requirement for significant private investment and the fact that NASA only pays for objective, demonstrated milestones ensures that the American taxpayer will receive exceptional value for the money."[24] He further noted, "I think it could be some of the best money NASA's ever spent."[25]

SpaceX will add the $278 million in NASA COTS funding to the millions it has spent (and will spend) to upgrade its Falcon 1 rocket into a Falcon 9. It also plans to develop the Dragon cargo and crew carriers that will ride on top of the Falcon 9. Depending on the mission, the Falcon 9 would be paired with either a Dragon cargo capsule or a Dragon crew capsule. The Falcon 9-Dragon design meets the COTS requirements for space vehicles that can service the International Space Station.

SpaceX has also received commercial contracts. For example, in May 2004, SpaceX received an order from Bigelow Aerospace for a 2009 launch of a Falcon 5 rocket, a larger five-engine version of SpaceX's Falcon 1, designed to carry Bigelow's inflatable habitats, discussed in Chapter 6, into space.[26]

Obstacles Faced by SpaceX

In addition to federal governmental regulation and licensing requirements, discussed in Chapter 7, startup companies, such as SpaceX, face other barriers. Musk turned to litigation, in an unsuccessful attempt, to break the Boeing-Lockheed duopoly on contracts with the U.S. government.

In May 2005, The Boeing Co. and Lockheed Martin Corp. announced plans to combine their heavy-lift Delta and Atlas rocket programs in joint venture, called the United Launch Alliance, because of decreased demand for commercial and government launches. The downturn was caused, in part, by the failure of a commercial market to materialize, forcing the companies to bear the high overhead costs associated with the development of and the infrastructure needed to support two separate rockets under the U.S. Air Force's heavy-payload Evolved Expendable Launch Vehicle (EELV) program.

Although Boeing and Lockheed claimed that their joint venture would save the U.S. government $100 to $150 million annually, by cutting the cost of putting military, spy, and research satellites into orbit, SpaceX counterattacked, filing a lawsuit in October 2005 alleging several antitrust violations. In addition to charging that the joint venture would be anti-competitive, the firm alleged a pattern of anticompetitive behavior and unfair business practices by Boeing and Lockheed intended to prevent SpaceX from entering part of the government launch market. Although SpaceX's Falcon 1 would not compete directly with rockets built by Boeing and Lockheed, SpaceX's future, larger model, Falcon 9, would compete with the rockets manufactured by the two aerospace giants.

In February 2006, a federal district court dismissed SpaceX's antitrust lawsuit for lack of standing to sue the two aerospace firms. Judge Florence-Marie Cooper held that "SpaceX's argument is utterly devoid of any concrete factual allegations regarding any type of actual injury suffered." Judge Cooper indicated that by its own assertions, SpaceX "is not yet ready to compete" with Boeing and Lockheed Martin to win government contracts under the EELV program, because its Falcon 9 was not scheduled for its maiden launch until late 2007. "Because it lacks such readiness [to compete with Boeing and Lockheed Martin's heavy launch rockets], its speculative claims regarding future harm are not yet ripe [for adjudication]," Judge Cooper reasoned.[27]

In October 2006, the Federal Trade Commission preliminarily approved the long-pending United Launch Alliance joint venture, creating a virtual monopoly in the heavy-payload area.[28] Both Boeing's Delta and Lockheed's Atlas rockets will continue to be produced, with the U.S. Air Force as the primary customer. Production will be consolidated at Boeing's Decatur, Alabama facility, with Lockheed's Denver office serving as the joint venture's headquarters and housing its engineering and administrative functions. Approval was given contingent on an FTC consent decree ordering the two companies to cooperate with potential competitors, such as Northrop Grumman Corp. and SpaceX.

Despite the adverse judicial ruling and the FTC's go-ahead to Boeing-Lockheed in May 2007, as a result of legislative changes, the future looked a little brighter for SpaceX and other upstart private space firms. Congress added language to the conference report accompanying the 2006 Fiscal Year Defense Appropriations Act directing U.S. military launch service contracts to "provide an annual opportunity for companies to present their qualifications to meet objective criteria of reliability, mission assurance, oversight, and cost credibility, and compete based on their ability to meet these criteria." Lawmakers mandated "the elimination of multi-year 'allocations,' 'pre-awards,' and 'block buys' from Buy-3 the third lot of EELV launches Buy-3 [the third lot of EELV launches] and future EELV launch services contracts."[29]

SpaceX's Key Personnel

Elon R. Musk was born in South Africa, and began his entrepreneurial career at age twelve, when he sold a computer code he developed for a "Space Invaders"-type video game, called "Blast Star," for $500.[30] He took that $500 and invested it in a pharmaceutical stock he had been following in Pretoria, his hometown, subsequently netting several thousand dollars. When he was seventeen he took advantage of his mother's Canadian citizenship and moved by himself to Canada. He then relocated to the United States to get his bachelor's degrees in physics at the University of Pennsylvania and economics at Penn's Wharton School. Determined to continue his American education, in 1995 Musk enrolled at Stanford to begin his doctoral studies in physics. Before attending any classes, he dropped out of Stanford to join the Internet boom that had engulfed Silicon Valley. "I could either watch it happen, or be part of it," Musk recalled.[31] He definitely became a part of it.

Before starting SpaceX, Musk built and sold two enormously successful Internet-based businesses. He developed a software program that allowed newspapers to post classified advertisements and local information on the Internet, called Zip2, and thus establish a web presence. In 1999, the company was sold to Compaq Computer Corp. for more than $300 million, of which his share was $22 million.

Instead of resting on his laurels, Musk invested some of his Zip2 profits in another business called X.com, which most notably allowed customers to transmit money over the Internet. The company he cofounded, later renamed its system PayPal, after acquiring a competitor that had originated the name. Only months after PayPal, Inc. went public in July 2002, eBay Inc. bought the firm, paying $1.5 billion, with Musk, PayPal's largest shareholder, receiving $150 million in eBay stock. By September 2002, *Fortune* placed Musk's net worth at $165 million, placing him at number 23 on a list of America's forty richest under the age of forty.[32] As a result of a rise in the price of eBay stock, Musk ranked number twelve on *Fortune's* 2004 list of the forty richest Americans under age forty, with an estimated net worth of $328 million.[33]

Musk and his former college roommate, Adeo Ressi, were driving down the Long Island Expressway in late 2000, trying to figure out what to do next in life. The tech bubble had burst. Musk planned to give over management of PayPal to someone with more business experience. Ressi and Musk looked into the darkness of the night. "There was a moment of silence," Ressi recalled. "I don't remember who said it, but someone said, 'Space.'"[34] After laughing, they discounted the idea, but it remained with Musk.

Musk clearly has a desire to leave a legacy. After hitting it big with PayPal, but before he cashed out, he asked a friend, "What is the most important thing that we can and should be doing?" Musk concluded, "Of all the great things humanity can do, experiencing stars is one of the greatest, if not the greatest."[35]

At age thirty-one, Musk, self-taught in aeronautics, decided to break into the commercial space market by starting Space Exploration Technologies where he serves as its chairman and chief executive officer. Too young to retire and too bored with the notion of creating another Internet company, he found a new ambition. "I like to be involved in things that change the world," Musk noted after he founded SpaceX. "The Internet did, and space will probably be more responsible for changing the world than anything else. If humanity can expand beyond the Earth, obviously that's where the future is."[36]

Despite his lofty rhetoric, Musk's SpaceX is starting small, with its aim of affordable launch vehicles for small payloads. Instead of figuring out a cheaper way to put tourists into space, Musk has focused on devising a way to lower launch costs generally, especially for the commercial satellite market. In deciding to attack launch cost in order to open up the U.S. space market, he noted, "Its peak of success came with the Apollo program in the late 1960s, and for the last 30 years it has moved sideways at best. One of the major reasons is launch cost."[37] Musk sees an excellent business opportunity in building "low-cost" rockets of his own design and manufacture that can put small satellites into orbit for some $6 million, compared to $30 million for NASA's cheapest rocket.[38]

Because of his enormous success in two prior ventures and the potential reliability of Musk's aerospace innovations, he has some observers in his corner. "Elon thinks bigger than just about anyone else I've ever met," asserted David Sacks, former chief operating officer at PayPal. "He sets lofty goals and sets out to achieve them with great speed." However, others are more skeptical. "He's a young, smart guy who has made a ton of money, but we have a saying in the launch business: The way to become a millionaire is to start with $1 billion," said Marshall Kaplan, former rocket engineer and current director of space programs for Strategic Insight, Ltd., a consulting firm based in Arlington, Virginia. And these odds make finding investors difficult. "It's going to be a highly risky and capital-intensive business at the same time," cautioned John Malloy, managing partner at Nokia Venture Partners in Menlo Park, California. "That's not a combination venture capitalists like to see."[39]

SpaceX's Technology

So far, SpaceX has been awarded nearly $500 million in advance contracts without even a completely successful launch to its name. This is likely because of the innovative design of Falcon 1, the company's first launch vehicle.

The 68-foot long, 28-ton Falcon 1 represents a streamlined version of expensive conventional launch vehicles used by the aerospace giants, Boeing and Lockheed Martin.[40] Musk opted for simplicity and, hopefully, increased reliability.

Falcon 1 is a two-stage, liquid oxygen and kerosene-powered rocket capable of firing a small payload of less than 1,000 pounds into low-Earth orbit in its basic configuration and one and a half tons with strap-on liquid boosters. As a two-stage vehicle, having only a one stage separation "event," with one engine per stage, it avoids the problems associated with complicated engine systems and multistage lift-offs. The Falcon 1 uses a new welding technology and sophisticated, lightweight materials. Its engine design uses only one fuel injector, not the traditional multiple fuel injectors. It utilizes computer chips, not circuit boards, thereby reducing wiring complications. In sum, designing the Falcon series, starting with Falcon 1, Musk sought simplicity because "it gets you both reliability and low cost."[41]

Musk achieves cost savings through Falcon 1's reusable first-stage and carefully monitoring expenditures, focusing on minimizing spending for the propulsion system, launch operations, and the general expenses of running a business. Falcon 1's first-stage is designed to drop by parachute back to earth and float for an ocean recovery by a salvage ship. Fishing the small first-stage out of the ocean is difficult, but it is cheaper than building a new rocket for each launch. The hope is that 80 percent of Falcon 1's components will be refurbished and used. However, even if recovered, it remains unclear how many times these components are reusable.

Musk remains vigilant about costs. In addition to Falcon 1's reusable first-stage, its design calls for engines, although less efficient and having lower performance goals, that are substantially less expensive to manufacture and operate. For Falcon 1, Musk designed and built his engines (and other key components, such as electronics, ground support equipment, guidance and control systems) as well as all primary structures in-house, rather than buying them from established, high cost suppliers. It purchases certain propulsion system elements, such as valves, and outsources some machining, plating, and coating work. After SpaceX constructs its Falcon rockets at its headquarters, they are then trucked (or shipped) to the launch pad, ready to go. X is not trapped, as Musk put it in "the high-cost culture of the space industry."[42]

SpaceX is also developing larger, more powerful launch vehicles, designed to carry people and larger satellites. Next up, is Falcon 5, with five engines, and Falcon 9, capable of carrying heavier loads into higher-level orbits with nine engines. With the Falcon 9, as noted earlier in this chapter, SpaceX would come into competition with Boeing's Delta 4 and Lockheed's Atlas 5 in the heavy-lift market, where the U.S. government is, at present, the main customer.

Falcon 9 will likely come in two versions, with the medium Falcon 9s, designed to carry 9,300 kilograms (20,502 pounds at low-Earth orbit) or 3,400 kilograms (7,495 pounds) to a geostationary transfer orbit, for a cost of $27 million per launch, and the heavier Falcon 9s, capable of orbiting 24,750 kilograms (54,564 pounds), for a cost of $78 million per launch.[43] The Falcon series of rockets, designed to lift light, medium, and heavy payloads, will be cost-effective because of either their first stage or both stages will be reusable. They will use similar engines, electronics, and guidance and control systems to maximize efficiencies and minimize costs.

For NASA's COTS program, SpaceX intends to use a capsule, called Dragon, atop its Falcon 9 rocket. Dragon entered development in 2004 and will be able to carry cargo or seven astronauts. Following his strategy of cutting costs, Musk will develop the thermal protection system, a key component, for the capsule in-house. However, SpaceX's COTS partner Paragon Space Development Corp. will, however, provide the life-support systems for the capsule.[44]

Financing SpaceX

To date, Musk's personal fortune has largely bankrolled SpaceX. Space entrepreneurs, such as Musk, Allen, Branson, and Bigelow, are spending their own funds, not seeking outside investors, because they can. "The people who've decided to do this have the means to do so," noted Musk. He continued, "They don't need to raise money."[45]

Musk's investment through the end of 2006, including work on the Falcon series of rockets and the Dragon capsule, totaled about more than $100 million. He plans to spend another $100 million to develop the Dragon cargo and crew carrier. Although Musk hinted that he would spend up the $300 million, he doubted that amount would be necessary, stating, "If need be, I will invest whatever it takes to make SpaceX the world leader in launch."[46]

SpaceShipOne, designed by Rutan and funded by Allen, winning the Ansari X Prize in October 2004 sparked investor interest in space tourism. Musk found it hard, but not impossible, to attract investors to SpaceX. "The investors would ask about prior successes in the field," he indicated. "In the launch area, there are none. But there are plenty of failures."[47] However, documents SpaceX filed with the U.S. Securities and Exchange Commission indicate that in 2002, SpaceX placed some $12 million worth of securities with five wealthy and/or high-income investors (so-called accredited investors), who may, however, have included Musk and his wife.[48] By the end of the first quarter of 2005, SpaceX placed an additional $10 million worth of securities with one accredited investor.[49]

Looking to a successful Falcon 1 operational launch, Musk anticipates getting not only more governmental and commercial orders but also venture capital funding. At some point, he may also attempt an initial public offering.

Along with Scaled Composites, SpaceX serves as a role model for other privately-held U.S. companies looking to commercial space activities. Ultimately, Musk has ambitious plans in mind, including the Falcon family of rockets to carry paying passengers to orbiting space hotels—such as Robert Bigelow is developing, as discussed in the next chapter—and tourists on eye-popping voyages around the Moon.

Although, at present, concentrating on bringing down the price of space launches and building a viable business based on governmental and commercial payload launch contracts, Musk evidences a far-reaching vision, beyond the scope of this book. "We have to do something dramatic to reduce the cost of getting to space," Musk stated. "If we can get the cost low, we can extend life to another planet." He candidly asserted, "I want to help make humanity a space-faring civilization."[50] Beyond building launch vehicles capable of providing low-cost, reliable access to space, making Mars reachable and habitable for humanity animates Musk's dream. "Mars has the possibility of being terraformed," Musk indicated, referring to the release of greenhouse gases on that planet, thereby trapping heat and creating a livable environment. "We could change the climate of Mars and make it quite Earth-like."[51] For Musk, "If we [SpaceX] can be one of the companies that makes it possible for humans to become a multi-planetary species, that would The Holy Grail. It sounds a bit crazy, but it's going to happen. . . . We're making progress toward a greater philosophical goal while building a sound business."[52]

Notes

1.   John Croft, "Changing the Low-Cost Launch Game," *Aerospace America* 42:3 (February 2004): 39-43.
2.   Brad Lemley, "Shooting the Moon," *Discover* 26:9 (September 2005) 28-34, at 32. See also Brian Deagon, "Will A Low-Cost Satellite Launch Firm Fly?" *Investor's Business Daily*, May 26, 2003, A6.
3.   Space Exploration Technologies Corp., Press Release, "SpaceX Selected for Responsive Space Launch Demonstration under DARPA Falcon Program," September 21, 2004.
4.   Mark Carreau, "Launch Could Lead to New Age of Tourism," *Houston Chronicle*, November 25, 2005, A1.
5.   Brian Deagon, "Space Pioneer Musk Betting $100 Million," *Investor's Business Daily*, November 17, 2004, A5. See also Mark Carreau, "Launch Could Lead to New Age of Tourism," *Houston Chronicle*, November 25, 2005, A1.
6.   Michael A. Dornheim, "First 30 Sec. Good. . .," *Aviation Week & Space Technology* 164:14 (April 3, 2006): 38-39, at 39.
7.   Brian Deagon, "SpaceX Looks To The Stars—And Fortunes Of Aerospace Giants," *Investor's Business Daily*, December 19, 2005, A5.
8.   Deagon, "Space Pioneer Musk." See also Brian Deagon, "Tech Entrepreneur Now Orbiting The Final Frontier—Space," *Investor's Business Daily*, May 2, 2005, A4.
9.   Tariq Malik, "Technical Problems Delay Falcon 1 Launch," *Space News* 17:6 (February 13, 2005): 15. See also Michael A. Dornheim, "SpaceX To Try Again," *Aviation Week & Space Technology* 164:3 (January 16, 2006): 419; Rob Coppinger, "SpaceX Fuel Tank Failure Hits Launcher's First Flight," *Flight International* 169:5017 January 3-9, 2006, 23; *Aviation Week & Space Technology*, "Space Exploration Technologies," 164:1 (January 2, 2006): 18; Peter Pae, "SpaceX's Maiden Rocket Launch Is Scrubbed Again," *Los Angeles Times*, December 20, 2005, C2; Deagon, "SpaceX Look to the Stars"; Peter Pae, "SpaceX Rocket Launch Is Reset," *Los Angeles Times*, November 29, 2005, C2; Peter Pae, "Hopes of Start-Up Rocket Company Are Riding on First Launch," *Los Angeles Times*, November 26, 2005, C1.
10.   Tariq Malik, "Technical Problems Delay Falcon 1 Launch," *Space News* 17:6 (February 13, 2005): 15.
11.   Jefferson Morris, "Trying Again," *Aviation Week & Space Technology* 165:5 (July 31, 2006): 40; *Houston Chronicle*, "El Segundo," July 19, 2006, A8; Brian Berger, "Pad Processing Error Doomed Falcon 1," *Space News* 17:14 (April 10, 2006):7; Michael A. Dornheim, "First 30 Sec. Good. . .," *Aviation Week & Space Technology* 164:14 (April 3, 2006); Michael A. Dornheim, "Musk," *Aerospace Daily & Defense Report* 217:61 (March 31, 2006): 4-5; Jefferson Morris, "Fuel Leak Brought Down Falcon 1, SpaceX CEO Says," *Aerospace Daily & Defense Report* 217:58 (March 28, 2006): 3; Andy Pasztor, "Entrepreneur's Rocket Suffers Setback During Maiden Launch," *Wall Street Journal*, March 27, 2006, A14.

12. Brian Deagon, "Commercial Space Industry Poised on Launching Pad?," *Investor's Business Daily*, September 25, 2006, A9.

13. Peter Pae, "Rocket Fails in SpaceX's First Launch," *Los Angeles Times*, March 25, 2006, C6 and Roger Fillion, "Rocket Entrepreneur to Forge Ahead," *Rocky Mountain News* (Denver), April 6, 2006, B2.

14. John Johnson Jr., "Private Rocket Hits New Heights," *Los Angeles Times*, March 21, 2007, A9; *Satellite News*, "Close Is Good Enough for SpaceX in Second Demonstration Launch," 30:12 (March 26, 2007)<Lexis-Nexis>.

15. Stan Crock, "The Final Frontier at Costco Prices," *Business Week* 3963 (December 12, 2005): 98, 100, at 100.

16. Roger Fillion, "Rocket Entrepreneur to Forge Ahead," *Rocky Mountain News* (Denver), April 6, 2006.

17. Stan Crock, "The Final Frontier at Costco Prices," *Business Week* 3963 (December 12, 2005). See also Brian Deagon, "All Systems Go for Space Entrepreneur Musk to Make History," *Investor's Business Daily*, November 25, 2005, A4.

18. *Space Daily*, "SpaceX Selected for Space Launch Demonstration under DARPA Falcon Program," September 22, 2004 <http://www.space-travel.com/reports>(December 7, 2006). See also Space Exploration Technologies Corp., Press Release, "SpaceX Selected for Responsive Space Launch Demonstration."

19. Space Exploration Technologies Corp., Press Release, "SpaceX Awarded $100 Million Contract from U.S. Air Force for Falcon I," May 2, 2005. See also Michael A. Dornheim, "Responsive Lift," *Aviation Week & Space Technology* 162:19 (May 9, 2005): 28; *Space Daily*, "SpaceX Awarded $100 Million Contract From USAF For Falson 1," May 4, 2005 <http://www.space-travel.com/reports>(December 7, 2006); *Inside the Air Force*, "SMC Awards Small Spacelift Contracts Totaling Up To $100 Million," 16:16 (April 22, 2005): 5; Rich Tuttle, "USAF picks Orbital, SpaceX for Responsive Small Spacelift," *Aerospace Daily & Defense Report* 214:15 (April 21, 2005): 5.

20. Andy Pasztor, "For Rocket Start-Up, Sky's the Limit, *Wall Street Journal Online*, September 15, 2005 <http://online.wsj. com/article-print/SB112674801818941381. html> (July 5, 2006).

21. Rocketplane-Kistler was formed from the March 2006 merger of Rocketplane Ltd. and Kistler Aerospace Corp., after the latter firm lost its financing, having spent some $600 million in trying to develop its reusable K-1 two-stage space vehicle. Kistler Aerospace went bankrupt before it could fly the K-1, which uses reliable, but inexpensive, Russian engines. Frank Morring, Jr. "COTS Competitors Say They'll More than Match NASA Funding," *Aerospace Daily & Defense Report* 219:43 (September 1, 2006): 4; Julie Bisbee, "Rocketplane Merges with Company," *Daily Oklahoman*, March 21, 2006 <2006 WLNR 4649283>; *Space Daily*, "Rocketplane and Kistler Announce Merge," March 9, 2006 <http://www.space-travel.com/reports>(December 7, 2006). For background on Kistler's K-1 vehicle see D. Fagan et al, "Kistler Reusable Vehicle Facility Design and Operational Approach," *Acta Astronautica* 52:7 (April 2003): 563-573; Greg Klerkx, *Lost In Space: The Fall of NASA and the Dream of a New Space Age* (New York: Pantheon, 2004), 115-141; G.E. Mueller and D. Kohrs, "The Aerospace Vehicle of the Future," in *The Space Transportation Market: Evolution Or Revolution* (Dordrecht, The Netherlands: Kluwer Academic, 2000).

22. I have drawn on National Aeronautics and Space Administration, Press Release, "NASA Selects Crew and Cargo Transportation to Orbit Partners," Release 06:295 August 18, 2006; National Aeronautics and Space Administration, "NASA Invests in Private Sector Space Flight with SpaceX, Rocketplane-Kistler, August 18, 2006;

Space Exploration Technologies Corp., Press Release, "SpaceX Wins NASA COTS Contract To Demonstrate Cargo Delivery To Space Station With Option For Crew Transport," August 18, 2006; Andy Pasztor, "NASA Selects Two Start-Ups to Build, Operate Spacecraft," *Wall Street Journal*, August 21, 2006, A2; Dave Ahern, "NASA Selects SpaceX, Rocketplane-Kistler For Cargo Orbiting Demo," *Defense Daily* 231:32 (August 21, 2006): 1. See also Frank Morring, Jr. "'Skin in the Game'," *Aviation Week & Space Technology* 165:14 (October 9, 2006): 66; Frank Mooring Jr., "COTS competitors"; Frank Morring, Jr., "Crew-Carrying Vehicle Work Started," *Aviation Week & Space Technology* 165:9 (September 4, 2006): 100.

23.   Brian Deagon, "Commercial Space Industry Poised on Launching Pad?," *Investor's Business Daily*, September 25, 2006, A9.

24.   Space Exploration Technologies Corp., Press Release, "SpaceX Wins NASA COTS Contract To Demonstrate Cargo Delivery To Space Station With Option For Crew Transport," August 18, 2006.

25.   Ian Watson, "Elon Musk Gets NASA Backing to Go Boldly," *The Business* (London), August 27, 2006 <http://www.thebusinessonline. com>(December 7, 2006).

26.   Tim Furniss, "Bigelow Signs as First Customer for SpaceX's Falcon V Variants," *Flight International* 165:4934 (May 18-24, 2004).

27.   The quotations in this paragraph are from Space Exploration Technologies Corp. v. The Boeing Co. and Lockheed Martin Corp., Order Granting Motions to Dismiss, CV 05-07533 FMC (MANx), United States District Court, Central District of California, February 16, 2006. See also Space Exploration Technologies Corp v. The Boeing Co. and Lockheed Martin Corp., Order Granting Motions to Dismiss, United States Distract Court, Central District of California, Order Granting Motions to Dismiss, CV 05-07533 FMC (MANx), May 15, 2006.

28.   Federal Trade Commission, Press Release, "FTC Intervenes in Formation of ULA Joint Venture by Boeing and Lockheed Martin," October 3, 2006; In the Matter of Lockheed Martin et al. FTC Decision and Order, Docket No. C-4188, May 1, 2007. See also Renae Merle, "Rocket Monopoly Approved," *Washington Post*, October 4, 2006, D1.

29.   Conference Report, *Making Appropriations For The Department Of Defense For The Fiscal Year Ending September 30, 2006, And For Other Purposes*, House of Representatives, 109th Congress, 1st Session, Report 109-359, December 18, 2005, 318. See also Michael Sirak, "U.S. Federal Court Dismisses Space X Lawsuit against Boeing and Lockheed Martin," *Defense Daily* 229:34 (February 22, 2006): 1.

30.   For background on Musk, I have drawn on Leslie Wayne, "A Bold Plan To Go Where Men Have Gone Before," *New York Times*, February 5, 2006, Section 3, 1; Josh Friedman, "Entrepreneur Tries His Midas Touch in Space," *Los Angeles Times*, April 22, 2003, Part 3, 1; Matt Krantz, "Taking a Cheap Shot at Space," *USA Today*, April 16, 2003, 6B; Seth Lubove, "Way Out There," *Forbes* 171:10 (May 12, 2003): 138-140.

31.   Matt Krantz, "Taking a Cheap Shot at Space," *USA Today*, April 16, 2003, 6B.

32.   Julia Boorstin, Jonah Freedman, Christopher Tkaczyk, "American's 40 richest under 40," *Fortune* 146:5 (September 16, 2002): 169-176, at 173.

33.   Telis Demos, Richard Morgan, Christopher Tkaczyk, "America's 40 richest under 40," *Fortune* 150:6 (September 20, 2004): 134-140, at 137.

34.   Michelle Kessler, "Star-Gazing Tech Titans Put Money where Dreams Are," *USA Today*, September 3, 2003, 1B.

35. *Ibid.*
36. Josh Friedman, "Entrepreneur Tries His Midas Touch in Space," *Los Angeles Times*, April 22, 2003, Part 3, 1.
37. *Ibid.*
38. Larry Claasen, "Sky Is No Limit for Rocket Man Elon Musk," *Business Day* (South Africa), October 22, 2003, 2.
39. Josh Friedman, "Entrepreneur Tries His Midas Touch in Space," *Los Angeles Times*, April 22, 2003, Part 3, 1.
40. For the technical details on Falcon 1, see Brad Lemley, "Shooting the Moon," *Discover* 26:9 (September 2005), 32-33; Craig Corault, "Bring It On!," *Aviation Week & Space Technology* 160:13 (March 29, 2004): 48-50, 52; John Croft, "Changing the Low-Cost Launch Game," *Aerospace America* 42:3 (February 2004): 40-41; Space Exploration Technologies Corp., Press Release, "SpaceX Performs First Rocket Engine Firing," March 19, 2003. See also Leslie Wayne, "A Bold Plan To Go Where Men Have Gone Before," *New York Times*, February 5, 2006, Section 3, 1.
41. Brad Lemley, "Shooting the Moon," *Discover* 26:9 (September 2005), 32.
42. Stan Crock, "The Final Frontier at Costco Prices," *Business Week* 3963 (December 12, 2005): 100. See also Frank Morring, Jr., "Second Tier," *Aviation Week & Space Technology* 165:23 (December 11, 2006): 52.
43. Rob Coppinger, "SpaceX Hopes to Supply ISS with New Falcon 9 Heavy Launcher," *Flight International* 168:5002 (September 13, 2005): 8. See also Crock, "Final Frontier," 10 and Wayne, "Bold Plan."
44. Rob Coppinger, "COTS Winners Give Teaming Details," *Flight International* 170:5051 (August 29-September 4, 2006): 27.
45. Brian Deagon, "Commercial Space Industry Poised on Launching Pad?," *Investor's Business Daily*, September 25, 2006, A9.
46. Michael A. Dornheim, "First 30 Sec. Good. . .," *Aviation Week & Space Technology* 164:14 (April 3, 2006): 39.
47. Brian Deagon, "Space Pioneer Musk Betting $100 Million," *Investor's Business Daily*, November 17, 2004, A5.
48. Space Exploration Technologies Corp., U.S. Securities and Exchange Commission, Form D, December 18, 2002.
49. Space Exploration Technologies Corp., U.S. Securities and Exchange Commission, Form D, March 9, 2005.
50. Leslie Wayne, "A Bold Plan To Go Where Men Have Gone Before," *New York Times*, February 5, 2006, Section 3, 1.
51. Seth Lubove, "Way Out There," *Forbes* 171:10 (May 12, 2003): 138.
52. Josh Friedman, "Entrepreneur Tries His Midas Touch in Space," *Los Angeles Times*, April 22, 2003, Part 3, 1. See also Stan Crock, "The Final Frontier at Costco Prices," *Business Week* 3963 (December 12, 2005): 100.

6

Developing Space Habitats:
Bigelow Aerospace Inc.

Bigelow Aerospace Inc., founded and headed by Robert T. Bigelow, a Las Vegas entrepreneur and owner of Budget Suites of America, is developing lightweight, inflatable space habitats, the first civilian space outposts. Bigelow has also created the $50 million America's Space Prize, to be awarded to the first privately-funded team to launch and return a spacecraft holding at least five people in two orbital flights, some 250 miles (400 kilometers) from Earth. By offering the prize, Bigelow seeks to encourage others to develop reusable a launch vehicle capable of docking with his orbiting space modules. He is focused on commercial human orbital, not suborbital, space utilization.

Beginning to Fulfill O'Neill's Dream

Bigelow's effort to develop space habitats provides a new dimension to the commercial spaceflight industry, building on the Gerard K. O'Neill's classic work, *The High Frontier*.[1] A Princeton University physics professor, O'Neill conceptualized that technology could solve many of humanity's problems, including the possibility of a rapidly diminishing lifestyle on Earth with the end of the petroleum era and an ever growing hunger for natural resources.

O'Neill developed a plan for self-sustaining, rotating space colonies designed to house thousands and ultimately millions of people, giving them Earth-like living conditions. His "Islands in Space" would be built at Lagrangian Point Five (L5), gravitationally stable, perpetually sunny locations between Earth and the Moon. These space cities would be constructed from materials mined on the Moon, hauled to L5 by robotic (or piloted) rockets, or launched from lunar mining sites in large batches by an electromagnetic catapult, a mass driver. The space cities, each one self-supporting through on-site agriculture and manufacturing, would generate solar power by means of large satellites that would be sent back to Earth. The importation of materials from Earth, such as hydrogen, could establish a basis for trade between our planet and O'Neill's space colonies.

Bigelow Aerospace Inc.

When Bigelow formed Bigelow Aerospace Inc. in 1999, he sought to more effectively utilize a traditional business model than NASA had in developing its TransHab Project, designed to provide living accommodations at the International Space Station and serve as a template for other space habitats. Bigelow licensed the intellectual property rights to the project after Congress nixed it in 2000, a victim of rising costs and the reluctance of Congress to give the International Space Station more funds.[2] "I've put together many, many projects involving a lot of money and a lot of people," Bigelow said. And unlike NASA, which is plagued by delays and skyrocketing costs, "I'm used to doing things pretty darn well on budget and pretty darn well on time."[3] Following this philosophy, Bigelow Aerospace's mission statement provides: "We intend to so reduce costs of station habitable structure as to make the difference between space stations being only government available or having space stations affordable for general business ownership."[4]

Bigelow expects to keep costs low, despite a $500 million commitment of his own funds into his space effort, by outsourcing rides on launch vehicles built by other companies (in part the inspiration for his $50 million America's Space Prize) and buying off-the-shelf hardware from private vendors, not high-priced defense contractors. Bigelow intends to save funds by paying companies, such as Space Exploration Technologies (see Chapter 5), Russian firms, Lockheed Martin, and ultimately the winner of America's Space Prize, to fly into space and to dock with his inflatable modules. Bigelow even figured out how to avoid paying fuel cost to maintain the habitat's orbital location. Following O'Neill's concept, the inflatable modules will be positioned at one of five LaGrange points between the Earth and the Moon, points which are not within the gravitational pull of any of the surrounding planets.[5]

Bigelow Aerospace plans to build its space habitat business looking to two types of customers: first, corporations that want to do research or manufacturing in space and second, a place where astronauts and others can perform research. For example, protein crystals grow to larger sizes when liberated from gravity; researchers find larger crystals easier to study, thereby simplifying drug design. The firm anticipates that pharmaceutical, biotechnology, and materials manufacturing companies will renew their interest in microgravity research, which they curtailed following the Columbia disaster in 2003. A space factory could take advantage of weightlessness to manufacture materials difficult (or impossible) to make on Earth. Corporations would lease a module for some period of time for business purposes. Bigelow's modules would also serve as space habitats for space pioneers, including astronauts from nations that lack access to the International Space Station or human spaceflight programs of their own as well as other researchers and explorers.[6] Because most tourists will want to spend several days in orbit, not just a few hours,[7] for space exploration to attain its full potential, a need exists for space hotels.

Thus far, Bigelow's personal outlays on his space venture run to about $75 million for building the Bigelow Aerospace physical plant in North Las Vegas, royalties on patents obtained from NASA, and building and testing prototypes of his modules. The biggest future outlays will center on building and then launching bigger prototypes and ultimately the actual modules.

America's Space Prize

In the fall of 2004, Bigelow created America's Space Prize to act as a catalyst for others developing spacecrafts to move along more quickly. The contest ends on January 10, 2010, by then Bigelow hopes to be well along on rolling out his space habitats to the public. As discussed in Chapter 2, it is also the year that NASA intends to retire its remaining space shuttles.

America's Space Prize will be awarded to a spacecraft that meets specifications set by Bigelow. In addition to reaching a minimum altitude of 400 kilometers (about 250 miles), the reusable craft must: carry no fewer than five people; demonstrate its ability to dock with a Bigelow Aerospace inflatable module; and be capable of remaining docked for at least six months. The craft must perform two consecutive safe and successful orbital missions within sixty days. It must also be a reusable space vehicle, with no more than 20 percent of the craft made from expendable hardware. The prize is limited to American contestants, who may not accept or utilize any government funding, including NASA's COTS program, or be owned by any government in any way.[8]

By offering the prize, Bigelow seeks to drive down the cost of orbital space travel, initially charging $14.9 million in 2012 for a trip to one of his habitats and a four-week stay. Bigelow's corporate counsel, Michael Gold, explained the motivation behind America's Space Prize. "I'm not going to pretend that we have purely no self-interest in this," Gold said. "We absolutely do. We need this system badly, and therefore whatever system we encourage under the prize with Mr. Bigelow's own money, we're going to make sure...can service the Bigelow Aerospace platform."[9] Bigelow put up half of the $50 million himself,[10] to spur the development of safe, reliable, low-cost launch capability.

Bigelow modeled his prize on the $10 million Ansari X Prize that Burt Rutan's SpaceShipOne, developed by Scaled Composites, won in October 2004, as discussed in Chapter 3, after traveling to suborbital space twice in less than a week. However, America's Space Prize will award five times more money than the Ansari X Prize. If successful, America's Prize winner will have developed the first commercial, manned orbital spacecraft that will survive a 17,500 mph reentry. In addition to the $50 million, the America's Prize winner will be guaranteed the first rights on a contract from Bigelow Aerospace for ongoing orbital servicing missions to the firm's space modules.[11]

Bigelow Aerospace's Key Personnel

Robert Bigelow, a self-made entrepreneur and hotel magnate, retained an interest in space while building a motel empire in the American Southwest.[12] Born and raised in Las Vegas, Bigelow graduated in 1967 from Arizona State University with a degree in real estate and banking. Throughout the 1970s and 1980s, following his father into real estate, Bigelow built his empire buying and then building apartment houses and motels around the Las Vegas area. He established Budget Suites of America in 1988, which today has sixteen motels in Nevada, Arizona, and Texas, and two Suites of America locations in Arizona and Texas, designed to attract travelers looking for an extended stay at reasonable rates.

As a child, Bigelow developed an interest in space and space colonization when he was eight years old, when his grandparents told him they had seen a UFO, glowing and radiating colors, coming at them in the Nevada desert. According to Bigelow, "They expected they would die,"[13] but his grandfather swerved off the highway and they watched out the car's back window as the object took off at a steep angle. When he was just fifteen years old, he made it his life's goal to become involved in space exploration. Realizing it would be expensive, he set out to make a fortune. In 1999, he established Bigelow Aerospace to make good on that promise to himself.[14] He hired veteran space-travel engineers to perfect the necessary technology. And always the businessman, Bigelow expects to make a profit while doing it. "With profit calling the shots, [the space tourism industry] is going to take off," Bigelow has said.[15] The firm aims to become profitable by 2010-2013.[16]

Bigelow also hopes that one day space travel will a provide spiritual, rather than a mere commercial, experience. When the spaceship passes the Moon, Bigelow envisions a special ceremony on the captain's bridge. "Because arriving at the moon is going to be an epiphany for most passengers," Bigelow mused, "a very, very profound moment."[17]

Bigelow, with estimated worth approaching one billion dollars as of 2000, invested in assorted non-profit ventures prior to founding Bigelow Aerospace. In 1996, he purchased a 480-acre ranch in Utah as a research base for the National Institute for Discovery Science, which he founded in 1995, having the stated mission of seeking evidence of UFOs and animal mutilations associated with UFO activity.[18] Before its 2002 termination and replacement by a scholarship program, in 1997 Bigelow pledged $3.7 million to the Consciousness Studies Program at the University of Nevada, Las Vegas, which included courses focused on the rational investigation of the mysteries of human awareness, including the possibility of the persistence of human consciousness after death. The program offered a limited number of university classes each semester, as well as workshops and public lectures by authorities in the fields of survival of bodily death and paranormal phenomena.[19]

Throughout his investments in paranormal phenomena, Bigelow kept quiet about his ultimate goal, building space habitats for commercial purposes. "I didn't even tell my wife," he stated. "She never knew. Because it's possible that that kind of dream would never happen."[20] Even after Bigelow began work to realize his dream, he rarely granted interviews or attracted any media attention until after he publicly announced the $50 million prize in the fall of 2004. However, he long believed in the global parameters of space privatization. "I'm a confirmed believer in the existence of UFOs," said Bigelow in 2000 when asked if his project was inspired by Buck Rogers. He continued, "[S]o how can I pooh pooh Buck Rogers stuff? This kind of stuff is not Buck Rogers. It's around the corner. The next century is not a competition between the United States and Russia anymore. China and Europe are very much in the space race. It's a whole new ball game that has begun. The European Space Agency, the Japanese and the Chinese are very serious competitors in the race to privatize space."[21]

Although Bigelow subsequently became more vocal about his project, he nevertheless realized the risks involved. "It's a gamble," he asserted. "It's a huge gamble."[22] He also recognized his project's dependence on the private launch system hopefully developed by the winner of America's Space Prize. As he stated, "We're building a stadium and if the people are not able to get there, the stadium will be useless." Alluding to the risks that arise from a dramatic endeavor such as space travel, he further noted, "One of the worst things you could have would be a fatal accident early on. You can kill 40,000 people a year in car accidents and it's not devastating to the auto industry."[23]

However, because the privatization of space is tied to national pride, Bigelow believed that any risks, economic and human, are worth the potential payoff. "It has been embarrassing to a lot of people in the aerospace industry and in private industry that the Russians have had to show America how to develop a free enterprise system in space," asserted Bigelow, referring to the Russian space agency's launches of wealthy individuals to the International Space Station for some $20 million each, discussed in Chapter 6. "We need to show not only that we can do it, but we can do it better," he maintained.[24]

Unlike Burt Rutan and his strained relationship with NASA, Bigelow has embraced a partnership with the often frustrating bureaucracy. His firm has an exclusive licensing agreement with NASA for two patents related to the agency's TransHab and radiation shielding technology, and another license for low impact docking system technology, with both exclusive and non-exclusive components.[25] Bigelow also entered into series of agreements, under the Space Act,[26] with NASA to work with former TransHab engineers still employed at the agency. "By virtue of these agreements we were able to get very valuable ongoing assistance and advice from NASA," stated Bigelow. "A lot of key NASA folks are in this [the Bigelow Aerospace] plant on a regular basis, and they help us out a lot."[27] He also retained as a consultant the senior NASA engineer who created the basic architecture for the module and who went on to an academic career after leaving the agency.

Bigelow has his criticisms of NASA, but they are often without the frustrated undertones of Rutan's jabs. "There are a vast number of people that are very depressed and disappointed over the miserable progress since 1970," Bigelow maintained: "NASA didn't have vision. They had a particular national security outlook. They didn't want Soviets to establish a moon base so they went to the moon, but they didn't have an encore. They didn't know what to do next. There was a complete lack of imagination from the senior leadership of NASA and the senior leadership of the White House."[28] Instead, Bigelow seems to see NASA as occupying a necessary, yet limited niche, in the space industry, and that its technology and brainpower are nevertheless essential to the development of a private market for space travel and exploration.

Bigelow Aerospace's Technology

Building lower-cost, more efficient space modules are what Bigelow is trying to do with Bigelow Aerospace. Bigelow's design is based on NASA's TransHab Project, which combined the words "transportation" and "habitat." As proposed, NASA would have launched the TransHab deflated in a space shuttle's cargo bay and once in space it would have inflated similar to a balloon to a three-story height.[29]

As envisioned by Bigelow, the TransHab, initially renamed the Nautilus, would be a lightweight, inflatable space module launched compressed, similar to a roll of paper towels, in the hull of a rocket.[30] After the vehicle's launch into orbit, "explosive bolts" release the girdle securing the hull and the habitat's life-support system inflates the structure, expanding it to between fifteen and twenty-two feet in diameter. Eventually, the module inflates to create three levels of working and living space, powered by solar panels that unfold from the bulkheads at each of its ends. Perhaps most impressively, the three levels of the inflated Nautilus structure are filled with breathable air, so astronauts and, eventually, researchers, employees, and space explorers, would be able to move around the structure without wearing spacesuits.

Inflatable, 45-foot by 22-foot Nautilus modules, each weighing twenty to twenty-five tons, could be docked together to form a small space station. With 330 cubic meters of volume, about the size of a three bedroom residence, each Nautilus module would have almost three times the volume of the individual modules that make up the International Space Station. Because the expandable inflatables can be packed into the nose of a rocket, they thus offer significant launch-weight savings, allowing the use of less-powerful launch vehicles.

Despite Bigelow Aerospace's close working relationship with NASA, Nautilus will differ considerably from NASA's proposed TransHab concept. After performing more than fifty ballistic impact tests, Bigelow Aerospace improved upon NASA's multilayered protection system, and has introduced lower cost materials, such as a graphite-fiber composite and layers of advanced materials, to protect the shell from space debris, among other hazards. Most notably perhaps, Nautilus will not be shaped like a truck tire, as TransHab was, but like a giant watermelon (or zeppelin). "Instead of TransHab's vertical layer cake, I turned the Nautilus design sideways so you do not have such short line-of-sight distances inside," Bigelow explained, "thereby making it more amenable for long-term living in space."[31]

Unfortunately, technological problems are not all that stand in the way of space pioneers putting and maintaining a habitat in orbit. Although the Commercial Space Launch Amendments Act of 2004,[32] discussed in Chapter 7, more than anything else, was designed to prevent space explorers from suing their carriers, it seems inevitable that the statute and the accompanying administrative regulations will be reviewed and expanded as the private sector space market grows in importance, bringing flights and hotel stays within the financial and practical reach of an ever increasing number of middle class individuals and families. That said, the U.S. Federal Aviation Administration (FAA) seems willing to support the private sector, at least in its initial efforts.

In May 2004, Bigelow's firm awarded launch contracts for its prototypes to SpaceX for its Falcon 5 rocket and to the Russian-Ukrainian commercial launch company Kosmotras, for its Dnepr rocket.[33] In the future, Bigelow may also use an improved version of Lockheed Martin's Atlas 5 launch vehicle.[34]

By November 2004, U.S. regulators cleared the first Genesis prototype for an orbital mission so that Bigelow Aerospace could begin to test the safety and reliability of its modules in space.[35] After reviewing the firm's application for the orbital launch of an inflatable habitat—first of it kind—for eight months, the FAA's Office of Commercial Space Transportation found that it met the U.S. government's standards for radiation exposure and construction materials, and cleared the Genesis inflatable module for launch by either a Falcon 5 rocket or a Dnepr rocket, the latter a commercial version of an intercontinental ballistic missile. The licensing of Genesis, a one-third scale prototype of the habitat, represents an important step for Bigelow Aerospace's project as well as an indicator of the U.S. government's effort to encourage private sector space ventures.

Bigelow scored a big triumph in 2006 with the successful deployment of Genesis 1, for which he received the space industry's annual Space Achievement Award. On July 12, 2006, a Dnepr rocket lofted a Genesis 1 module into orbit, 348 miles (550 kilometers) above the Earth, from a missile base in Siberia. Computer-controlled air pressure tanks keep the prototype inflated, with Kevlar membranes restraining inflation and eight solar panels providing electricity.[36] Over the next five years engineers and scientists will study how Genesis 1 withstands space radiation, space debris, and micrometeoroids, some of the deadliest concerns for a space habitat. It is expected to orbit for about thirteen years before burning up in the atmosphere. In commenting on the launch of the Genesis 1 module, which when inflated has the silhouette of a ten-foot-long by eight-foot-wide sausage, George T. Whitesides, executive director of the National Space Society, a nonprofit space-advocacy group, called the launch "incredibly significant. This is the only real, funded project that's trying to create a destination in space privately, as opposed to the other folks, who are creating private launch vehicles."[37]

Bigelow launched a Genesis 2 inflatable in June 2007. Similar to Genesis 1, Genesis 2 will test a variety of items, including electronics, seal technology, and oxygen tanks. It will also carry more cameras. The Galaxy prototype, scheduled for launch in late 2008, will be twice as large as Genesis and one half the scale required for the operational module. It will provide tests of life-support system components and electronics; more exterior payload will be mounted on the Galaxy as well as more advanced solar arrays. Scheduled for launch in 2010 is the ten-ton Sundancer prototype, which will inflate to 180 cubic meters of volume. Tests with Sundancer will lead to the launch of the full-scale BA-330 module, a twenty-ton 330 cubic-meter inflatable, which could house up to six people.[38] After test of the BA-330 module in orbit, soon thereafter, a full-fledged Bigelow space habitat will be open for business. His inflatable pods will serve as the building block for space facilities, whether for research, manufacturing, or accommodations.

Despite all the potential perils of the private space exploration industry, Bigelow has embraced it. "[I'm] taking a huge risk, but it's fun. I love the subject. I like to keep it fun," he stated. "By the time you struggle with business, raise a family, and you manage to get to a point when there is some time available, you say to yourself, 'Aha, some day, if you're still alive, you are going to do something...' I would like to think we could look back and think we made a difference."[39] It is this entrepreneurial spirit and passion for space that makes Bigelow the perfect person for developing expandable space habitats. It makes Bigelow Aerospace a firm to watch as private sector space ventures take off.

Notes

1.   Gerard K. O'Neill, *The High Frontier: Human Colonies In Space*, Third Edition (Burlington, Ontario, Canada: Apogee, 2000). See also Giancarlo Genta and Michael Rycroft, *Space, the Final Frontier?* (New York: Cambridge University Press, 2003), 121-127.

2.   National Aeronautics and Space Administration Authorization Act of 2000, Public Law 106-391, Section 127 and Conference Report, National Aeronautics and Space Administration Authorization Act of 2000, House of Representatives, 106th Congress, 2d Session, Report 106-843, September 12, 2000, 31-32. See also Craig Covault, "Inflation Factor," *Aviation Week & Space Technology* 161:1 (July 5, 2004): 20-21, at 20; George Knapp, "The Ultimate Public-Private Partnership, *Las Vegas Mercury*, July 8, 2004 <http://stephensmedia.printthis> (February 8, 2007).

3.   Michael Belfiore, "The Five-Billion-Star Hotel," *Popular Science* 266:3 (March 2005): 50-57, 87, at 55.

4.   Bigelow Aerospace, Inc., "Mission Statement" <http:// bigelowaerospace. com>(May 12, 2005).

5.   John M. Glionna, "Space-Tourism's Hot Ticket," *Los Angeles Times*, May 22, 2000, A1.

6.   Taylor Dinerman, "The Bearable Lightness of Being," *Wall Street Journal*, (April 26, 2007): D7; Leonard David, "Bigelow Aerospace Aims for an International Market," *Space.com*, April 10, 2007 <http://www.space.com/news/070410_nss_bigelow.html> (May 24, 2007); Marc Kaufman, "Private Space of the Future," *Washington Post*, April 12, 2007, D1; Jeff Foust, "Big Plans, Low Prices," *The Space Review*, April 16, 2007 <http://www. thespacereview.com/article/85211> (May 24, 2007). For technical details on space hotels, see Buzz Aldrin and Ron Jones, "Changing the Space Paradigm: Space Tourism and the Future of Space Travel," in *Space: The Free Market Frontier*, ed. Edward L. Hudgins (Washington, DC: CATO Institute, 2002), 184-186. Patrick Collins, "Space Hotels: Civil Engineering's New Frontier," *Journal of Aerospace Engineering* 15:1 (January 2002): 10-19, analyzes the prospects for the development of space hotels, concluding that it could become a new business field.

7.   P. Collins, R. Stockmans, M. Maita, "Demand For Space Tourism In America and Japan, and Its Implications For Future Space Activities," *Advances in the Astronautical Sciences* 91 (1995): 601-610, at 604-605.

8.   Bigelow Aerospace, Inc., "America's Space Prize: Ten Primary Rules of the Competition" <http:bigelowaerospace.com/ space_prize.htm> (April 14, 2006).

9.   Jefferson Morris, "Bigelow Drawing Up Rules for Orbital Prize Competition," *Aerospace Daily & Defense Report* 212:20 (October 28, 2004): 3.

10.   Craig Covault, "Bigelow's Gamble," *Aviation Week & Space Technology* 161:12 (September 27, 2004): 54-58, at 54.

11.   *Ibid.*

12.   I have drawn on John Johnson Jr., "His Inn Will Be Way Out," *Los Angeles Times*, August 30, 2006, A1; Michael Belfiore, "The Five-Billion-Star Hotel," *Popular Science* 266:3 (March 2005); John M. Glionna, "Space-Tourism's Hot Ticket," *Los Angeles Times*, May 22, 2000, A1; Robert Hodierne, "Space Oddity," *Washington Post Magazine*, December 9, 2001, W14-W18, W-23, at W18, W23; Keith Rogers, "Race toward Space Tourism Interests LV Millionaire," *Las Vegas Review-Journal* (Nevada), October 11, 2004, 1B.

13.   John Johnson Jr., "His Inn Will Be Way Out," *Los Angeles Times*, August 30, 2006, A1.

14.   Michael Belfiore, "The Five-Billion-Star Hotel," *Popular Science* 266:3 (March 2005).

15.   John M. Glionna, "Space-Tourism's Hot Ticket," *Los Angeles Times*, May 22, 2000, A1.

16.   Craig Covault, "Commercial Inflation," *Aviation Week & Space Technology* 165:23 (December 11, 2006): 50-51.

17.   John M. Glionna, "Space-Tourism's Hot Ticket," *Los Angeles Times*, May 22, 2000, A1.

18.   Keith Rogers, "Race toward Space Tourism Interests LV Millionaire," *Las Vegas Review-Journal*, October 11, 2004, B1. See also National Institute for Discovery Science, "Mission Statement" <http://www.nidsci.org/mission.php>(December 13, 2006).

19.   Natalie Patton, "UNLV unplugs program on human consciousness," *Las Vegas Review-Journal*, November 8, 2002, B1 and Natalie Patton, "Mind Frontiers," *Las Vegas Review Journal*, April 15, 1987, B1.

20.   Michael Belfiore, "The Five-Billion-Star Hotel," *Popular Science* 266:3 (March 2005): 54.

21.   *Sunday Herald Sun* (Melbourne, Australia), "Hotel Out of this World," February 13, 2000, V6.

22.   Michael Belfiore, "The Five-Billion-Star Hotel," *Popular Science* 266:3 (March 2005): 52.

23.   Robert Macy, "Mogul Gambles $500 Million on Lunar Lodging," *Toronto Star*, January 29, 2000, 1.

24.   Judith Graham, "Critics," *Chicago Tribune*, April 24, 2002, N1.

25.   Exclusive License Agreement between the United States of America Represented by the National Aeronautics and Space Administration and Bigelow Development Aerospace Division, LLC, Exclusive Patent License Agreement No.: DE-362, n.d.; Exclusive License Agreement between the United States of America Represented by the National Aeronautics and Space Administration and Bigelow Development Aerospace Division, LLC, Patent License Agreement No.: DE-371, n.d.; Exclusive License Agreement between the United States of America Represented by the National Aeronautics and Space Administration and Bigelow Development Aerospace Division, LLC, Partially Exclusive Patent License Agreement No.: DE-385, n.d. See also National Aeronautics and Space Administration, "Bigelow Aerospace Continues Relationship with NASA-JSC for Space Habitat Technology and Private Sector Space Development <http:technology.jsc.nasa.gov/bigelow_story. cfm>(November 27, 2006) and Leonard David, "Bigelow Aerospace to Tackle Inflatable Space Habitats," Space.com, May 24, 2004 <http://www.space.com/news/businessmonday_040524.html> (November 27, 2006).

26.   Nonreimbursable Space Act Agreement between the National Aeronautics and Space Administration Lyndon B. Johnson Space Center and Bigelow Aerospace to Support Development of Inflatable Space Structures, April 4, 2002; Nonreim-

bursable Space Act Agreement between the National Aeronautics and Space Administration Lyndon B. Johnson Space Center and Bigelow Aerospace to Support the Development of Expandable Space Structures, October 19, 2004; Space Act Agreement between the National Aeronautics and Space Administration Lyndon B. Johnson Space Center and Bigelow Aerospace to Support the Development of Expandable Space Structures, January 27, 2005; Reimbursable Agreement between the National Aeronautics and Space Administration Lyndon B. Johnson Space Center and Bigelow Development Aerospace Division, LLC to Support the Development of Expandable Space Structure, August 5, 2005; Space Act Agreement Between NASA Johnson Space Center and Bigelow Development Aerospace Division, LLC on Expandable Space Structures, n.d. The National Aeronautics and Space Act of 1958, as amended, specifically, 42 USC §2473(c)(5), permits NASA to enter into Space Act Agreements. This type of agreement is a flexible arrangement allowing NASA to work cooperatively with the private sector to facilitate the commercial development of space and support of NASA's mission and the U.S. national space priorities. As a collaborative research and development effort, a Space Act Agreement may provide for an ongoing exchange of personnel as well as the use of NASA facilities, expertise, equipment and technology. National Aeronautics and Space Administration, Space Act Agreements Manual, NPR 1050.1.

27.   Craig Covault, "Bigelow's Gamble," *Aviation Week & Space Technology* 161:12 (September 27, 2004): 56-57.

28.   Julian Borger, "2001," *The Guardian* (London), January 6, 2001, Weekend Section, 6.

29.   Craig Covault, "Mars Initiative Leads Station Course Change," *Aviation Week & Space Technology* 147:23 (December 8, 1997): 39-40.

30.   I have drawn on Michael Belfiore, "The Five-Billion-Star Hotel," *Popular Science* 266:3 (March 2005); and Craig Covault, "Inflation Factor," *Aviation Week & Space Technology* 161:1 (July 5, 2004): 20.

31.   Craig Covault, "Inflation Factor," *Aviation Week & Space Technology* 161:1 (July 5, 2004): 21.

32.   Public Law 108-492.

33.   Tim Furniss, "Bigelow signs as first customer for SpaceX's Falcon V variant," *Flight International* 165:4934 (May 18-24, 2004): 38.

34.   *Aviation Week & Space Technology*, "In a major development," 165:12 (September 25, 2006): 22; *Space Daily*, "Bigelow And Lockheed To Study Using Atlas 5 For Manned Launches," September 25, 2006 <http:www.space-travel.com/reports>(February 8, 2007); Rob Coppinger, "Atlas V may be Bigelow launcher," *Flight International* 170:5060 (October 31 - November 6, 2006): 25.

35.   *Flight International*, "Inflatable Module Cleared to Fly," 166:4963 (December 7-13, 2004): 31 and Frank Mooring Jr., "Inflatable Milestone," *Aviation Week & Space Technology* 161:21 (November 29, 2004): 23.

36.   *Aviation Week & Space Technology*, "Bigelow Aerospace Genesis I," 165:4 (July 24, 2006): 18; *Space Daily*, "Bigelow Releases First Images Inside Genesis," July 25, 2006 <http://www.space-travel.com/reports>(February 8, 2007); *Aerospace Daily & Defense Report*, "Bigelow Inflatable Station Module Undergoes Tests Following Launch," 219:7 (July 13, 2006): 1.

37.   Peter N. Spotts, "Business Takes One Small Step into Space," *Christian Science Monitor*, July 17, 2006, 3.

38.   Craig Covault, "Commercial Inflation," *Aviation Week & Space Technology* 165:23 (December 11, 2006): 50-51.

39.   Julian Borger, "2001," *The Guardian* (London), January 6, 2001, Weekend Section, 6.

7

The Legal Environment for Private Sector Space Enterprises

Legal barriers, both transnational and domestic, exist with respect to the development of private sector space enterprises. These barriers limit the private sector's potential by creating uncertainty, adding financial risks, and forcing ventures to follow expensive and time-consuming bureaucratic processes. It is especially important to note that space exploration and travel is already a risky endeavor, technologically and financially. Thus, when you add legal uncertainty and bureaucratic barriers to the mix, it becomes more difficult to find willing investors. Nevertheless, as discussed in Chapters 3 through 6, a new crop of entrepreneurs has popped up in the United States, hoping to turn a profit on everything from cheaper cargo launch systems to space exploration and even celestial habitats.

After briefly introducing the international legal regime for outer space and U.S. laws implementing the jurisdiction conferred by transnational law, this chapter analyzes: U.S. licensing and regulatory requirements; liability considerations under transnational and U.S. law; and property rights and resource appropriation under transnational law.

As discussed in this chapter, the private exploration of and settlement in outer space is legal, both under United States and international laws. However, the uncertainty of transnational legal standards regarding property rights in outer space, nevertheless, serves as a disincentive to private sector space enterprises. Business entities and investors, unsure of their rights and lacking assurance that their efforts and investments will receive legal protection, may hesitate to undertake the risks involved in developing new technologies and investing financial and human resources if they cannot be assured of some reasonable return. Similar to any other industry, the commercial space market needs a well-defined and minimally-invasive domestic and transnational legal and administrative regime in order to reach its full potential. Because profit provides an excellent motivator for businesses, a need exists for legal certainty on the transnational level, which would provide an incentive for private sector investment in outer space development. The 1967 Outer Space Treaty, to which we turn, must be updated to provide property rights for commercial entities in outer space, including the Moon, other planets in the solar system, and various asteroids. Also, the Moon Treaty to which no space-faring nation is a signatory, ought to be abrogated.

Transnational Legal Regime For Outer Space: An Overview

The 1967 Treaty on Principles Governing the Activities of States in the Exploration and Use of Outer Space, Including the Moon and Other Celestial Bodies, commonly known as the Outer Space Treaty,[1] is the first and most significant transnational space treaty. Other transnational space treaties, considered in this chapter, are: 1968 Agreement on the Rescue of Astronauts, the Return of Astronauts and the Return of Objects Launched into Outer Space (Rescue Agreement);[2] 1972 Convention on International Liability for Damage Caused by Space Objects (Liability Convention);[3] 1976 Convention on the Registration of Objects Launched into Outer Space (Registration Convention);[4] and 1984 Agreement Governing the Activities of States on the Moon and Other Celestial Bodies (Moon Agreement).[5] More than one hundred nations, including the United States, have signed the 1967 Outer Space Treaty, far more than any other space-related treaty.[6] Thus, all space activities, whether public or private, must be considered in light of the Outer Space Treaty.

Despite its name, the Outer Space Treaty does not define where air space ends and outer space begins. The Preamble to the treaty recognizes "the common interest of all mankind in the progress of the exploration and use of outer space for peaceful purposes,"[7] but it fails to define the term "outer space." According to Article I, the 1967 treaty applies to "outer space, including the Moon and other celestial bodies," but does not state what the term "outer space" includes.[8] Possible definitions could include low-Earth orbit, orbital space, or any national airspace once one reaches a certain altitude.

The ambiguity over the extent of "outer space" under the 1967 treaty came to a head with the 1976 Bogota Declaration,[9] where a group of eight equatorial nations sought to claim sovereignty over the "slices" of space directly over their respective national boundaries to the geostationary orbit, located 22,300 miles (about 35,785 kilometers) above the Equator. Geostationary (or geosynchronous) orbiting vehicles match the earth's rotation, thereby remaining in a stationary position above the Earth's surface. The eight nations claimed that because this orbit is a physical fact arising from the Earth's nature, it should not be considered part of outer space. This attempt to expand national airspace into orbital space proved unsuccessful, however, as it was seen to contradict the terms of the Outer Space Treaty. As a result, the space-faring signers of the 1967 treaty ignored the 1976 declaration, which sought to give those equatorial states jurisdiction over the lucrative market for geosynchronous communications satellites.

Leaving unresolved the definition of "outer space," for purposes of this introductory overview of the Outer Space Treaty, three points are of significance: first, the treaty is designed to prevent conflicts in outer space by banning weapons of mass destruction; second, the treaty indicates that signatory nations should provide assistance to astronauts; and third, nations which are parties to the treaty are responsible for space activities carried out under their auspices. The Outer Space Treaty prohibits any signatory nation from placing in orbit around the Earth "any objects carrying nuclear weapons or any other kinds of weapons of mass destruction...," from installing such weapons on celestial bodies, or using space to station such weapons in any way.[10] Furthermore, no nation is permitted to build military bases, test any type of weapon, or engage in any military maneuvers in space. The use of military personnel for scientific purposes, however, is permitted by the treaty. The emphasis on the peaceful use and settlement of space represents an outgrowth of the political climate in which it was drafted. When the treaty was signed, the world was in the midst of the Cold War. Both the United States and the former Soviet Union, the only space-faring nations at the time, wanted assurances that their respective enemy would not use outer space as a military outpost.

Second, emphasizing the peaceful, universalistic tone of the treaty, it further states that the astronauts from any nation should be regarded as "envoys of mankind in outer space,"[11] and that each signatory nation should provide the astronauts with emergency assistance if there ever were an accident, distress, or emergency landing on another nation's territory or in the oceans. This approach extends to the astronauts in space, as the treaty asks nations to "immediately inform the other States Parties to the Treaty or the Secretary-General of the United Nations of any phenomena they discover in outer space, including the moon and the other celestial bodies, which would constitute a danger to the life or health of the astronauts."[12] Furthermore, the 1968 Rescue Agreement provides that personnel aboard spacecraft who have suffered injury, are in distress, or have landed in another country should be rescued, rendered "all necessary assistance," and "safely and promptly returned to representatives of the launching authority."[13] Space explorers may constitute spacecraft personnel bringing them within the obligations of this agreement. The broad mandate of the Rescue Agreement is not, however, accompanied by any specific guidelines and is silent on a nation's financial obligation for a rescue mission.

Third, signatories to the 1967 Treaty bear international responsibility for all their activities in outer space whether "carried on by governmental agencies or by non-governmental entities...."[14] Nations party to the Treaty must authorize and continue to supervise the activities of their respective entities in outer space. In other words, private sector entities may only act under the aegis of a nation which assumes responsibility and, therefore, as discussed later in this chapter, the accompanying liability for their actions.

Nations continue to exercise jurisdiction over their respective objects and individuals once in outer space. The 1967 treaty provides that nations which are party to the treaty "on whose registry an object launched into outer space is carried shall retain jurisdiction and control over such object, and over any personnel thereof, while in outer space or on a celestial body."[15]

The issue of each nation's jurisdiction is dealt with under transnational law through a system of registration. The 1976 Registration Convention requires a launching nation to maintain a registry of launched space objects. The convention provides, "When a space object is launched into earth orbit or beyond, the launching State shall register the space object by means of an entry in an appropriate registry which it shall maintain."[16] Although the convention is inapplicable to suborbital vehicles which are not launched into the Earth's orbit, the registration system provides a means for the identification of space objects that may cause damage.

Because nations exercise jurisdiction over spacecraft registered with them, the nation of registry can subject its space objects, including those of private sector entities and personnel, to its national laws so long as these domestic rules and regulations do not conflict with transnational law. The United States may thus legislate with respect to a broad range of private sector space activities.

United States Law and Regulations Regarding Private Sector Space Activities

For purposes of this book, United States laws and administrative regulations regarding commercial space ventures consist of three aspects: legislation implementing the jurisdiction conferred by the Outer Space Treaty; licensing and regulatory requirements; and liability considerations. This section considers each of these aspects.

Examples of United States Legislation Implementing The Jurisdiction Conferred by the Outer Space Treaty

The United States had enacted numerous laws, particularly dealing with taxation and intellectual property, pursuant to the grant of authority contained in the Outer Space Treaty. Following the general American worldwide sourcing rule for the imposition of U.S. income taxes, the Internal Revenue Code provides that "any income derived from a space . . . activity - (A) if by a United States person, shall be sourced in the United States."[17]

As to intellectual property, through the 1990 Patents in Outer Space Act, the United States extended the reach of its patent legislation to inventions made onboard U.S.-registered space objects. This act provides, in part, that "(a) Any invention made, used or sold in outer space on a space object or component thereof under the jurisdiction or control of the United States shall be considered to be made, used or sold within the United States," subject to certain exceptions.[18]

Other U.S. legislation implements the jurisdiction conferred on nation-states by the 1967 Treaty. For example, the maritime and territorial jurisdiction of the United States for federal criminal law purposes extends to "[a]ny vehicle used or designed for flight or navigation in space on the registry of the United States pursuant to [the Outer Space Treaty], while that vehicle is in flight. . . ."[19]

Licensing and Regulatory Requirements

Congress has enacted several laws to facilitate commercial space launch services, beginning with the 1984 Commercial Space Launch Act (the 1984 Act),[20] and then, in subsequent legislation discussed in this section, the 1988 Commercial Space Launch Act Amendments, the 1998 Commercial Space Act, and finally the 2004 Commercial Space Launch Act Amendments. The 1984 Act, as subsequently amended, provides the groundwork for the private sector taking space exploration and payloads from NASA's grip. Before 1984, the federal government served as the sole provider of space launch services in the United States. The federal government either provided its own vehicles or contracted for unmanned space launch vehicles from private firms.

Beginning in July 1982, however, National Security Decision Directive (NSDD) 42[21] marked the first time in the history of the U.S. space program that a high-level official federal government document directly referred to the American private sector. With NSDD 42 the Reagan administration began to challenge and then overcome the policy that the U.S. government would provide the delivery of all space services. Goal four of the NSDD 42 specifically focused on "expand[ing] United States private-sector investment and involvement in civil space and space-related activities."[22] Another principle of the directive was that "[t]he United States encourages domestic commercial exploration of space capabilities, technology, and systems for national economic benefits."[23] The directive also declared: "The United States government will provide a climate conducive to expanded private sector investment and involvement in civil space activities, with the due regard to public safety and national security."[24]

The 1984 Act codified most of the provisions of previously-issued Executive Order 12465,[25] which had designated the U.S. Department of Transportation as the "lead agency within the Federal Government for encouraging and facilitating commercial [expendable launch vehicles (ELVs)] activities by the United States private sector." The executive order authorized the Department of Transportation to "expedite the processing of private sector requests to obtain licenses" to operate ELVs. With commercial ELVs in mind, the 1984 Act only referred to and sought to regulate launches, launch vehicles, and launch sites.[26]

By establishing the Department of Transportation as the lead federal agency to oversee and coordinate commercial space launch activities in the United States, the 1984 Act performed three basic functions. First, it authorized the Department of Transportation to license the launch of vehicles as well as the operation of launch sites.[27] Second, the act gave the department authority to regulate the commercial space transportation industry, to the extent necessary, to ensure compliance with the international obligation of the United States, and to protect public health and safety, the safety of property, and the national security and foreign policy interests.[28] Third, it also gave the department authority to encourage, facilitate, and promote commercial space launches by the domestic private sector by simplifying and expediting the issuance of launch licenses.[29]

In an effort to bring within one federal agency the private sector launch licensing process, the Office of Commercial Space Transportation (OCST) was created in the Department of Transportation in 1984. By becoming the focal federal agency for space commercialization, the OCST assumed the responsibility for regulating launch companies.

In 1988, Congress amended the 1984 Act. The Commercial Space Launch Act Amendments of 1988 established a new liability risk-sharing regime, which remains in effect as the result of periodic extensions of the law.[30] While the 1984 Act required a launch licensee to obtain liability insurance in an amount considered necessary by the Department of Transportation, the 1988 legislation required launch providers to obtain specified amounts of liability insurance or at least the maximum available on the world market at a reasonable cost.

In 1995, the OCST became the Office of the Associate Administrator for Commercial Space Transportation (AST) and the agency was transferred to the Federal Aviation Administration (FAA or FAA/AST), which received launch regulatory authority. Thereafter, the Commercial Space Act of 1998[31] extended the FAA/AST's regulatory authority to reentry vehicle operations. The 1998 act sought to remove the existing barriers to private firms in the space market by removing the previous ban restricting private enterprises from bringing back humans, payloads, and reentry vehicles from outer space. Newly empowered, the FAA/AST then implemented licensing procedures for commercial, reusable space vehicles to launch and land in the United States.

Until December 2004, several federal agencies jousted for jurisdiction over suborbital spacecraft. These craft enter space using rocket power, thus appearing to be under the purview of the FAA/AST. However, several of these craft behave similar to airplanes for take-off and landing, thereby seeming to be subject to the FAA's Regulation and Certification Group (RCG), which regulates experimental aircraft. Because the high cost of complying with either the AST or RCG (or both) systems threatened to destroy the nascent industry, suborbital spacecraft needed its own regulatory regime.

The Commercial Space Launch Amendments Act of 2004 (the 2004 Act)[32] seeks to define human flight as a commercial activity distinct from the aviation industry and streamline the regulatory barriers to certain launches, while balancing safety with innovation. Thus, the findings section of the act states: "the United States should encourage private sector launches, reentries, and associated services and, only to the extent necessary, regulate those launches, reentries, and services to ensure compliance with international obligations of the United States and to protect the public health and safety, safety of property, and national security and foreign policy interests of the United States."[33]

Designed to modernize the regulatory environment for commercial spaceflight, the act passed in the House of Representatives in November 2004 and the Senate in December 2004.[34] The act amends the "Findings and Purposes" section of the law to provide that "a critical area for the Department of Transportation is to regulate the operations and safety of the emerging commercial human spaceflight industry. . . ."[35] Thus, the act gives the Secretary of Transportation discretion to delegate authority over human spaceflight, subject to a one-stop shopping requirement that mandates the issuance of one license (or permit), to an entity within the department, currently, the FAA.

Although an extended analysis of current U.S. laws and regulations regarding private sector spaceflight is beyond the scope of this book, an observation is in order. With the growth of commercial space activities, the existing regime will come under review to see if it imposes too burdensome and costly licensing and regulatory requirements. Excessive regulation in the name of safety, whether safety of crew, passengers, or the general public, slows innovation and imposes needless costs and time delays on investors and business entities. For purposes of this section it is important to distinguish rules pertaining to the licensing of manned spacecraft carrying passengers or a payload for compensation from those pertaining to an experimental permit for test flights of manned suborbital spacecraft not carrying passengers or payload for compensation.

Licensing of Manned Spacecraft Carrying Passengers or a Payload For Compensation

For a manned spacecraft to carry passengers or a payload for compensation, a license, not an experiment launch permit, is required. In addition to licensing commercial launches that take place in the United States, the FAA licenses all overseas launches by U.S. citizens or companies, subject to certain exceptions.[36] The licensing process is complex, time-consuming, and costly.

In brief, the FAA's licensing process includes an application evaluation review made up of the following major parts: a policy review, a payload review, a safety evaluation, a financial responsibility determination, and an environment review.[37] The FAA reviews a license application to determine if it presents any issues impacting national security or foreign policy interests or international obligations of the United States. Under the payload review, for any non-U.S. government payload whose reentry is not subject to review by another federal agency, the FAA determines whether an applicant has obtained all required licenses and authorizations. It reviews a payload proposed for launch and reentry to determine whether it would jeopardize public health and safety, safety of property, national security or foreign policy interests, or international obligations of the United States. The safety review focuses on whether an applicant can safely conduct its proposed activities, without jeopardizing public health and safety, and includes both quantitative and qualitative analyses. An applicant typically demonstrates financial responsibility, a topic discussed later in this chapter, to compensate for the maximum probable loss by purchasing liability insurance. Because the issuance of the license is considered to be a major federal action under the National Environmental Policy Act (NEPA),[38] the environmental evaluation considers whether the proposed space transportation activity would pose an unacceptable danger to the environment. An applicant must provide sufficient information to enable the FAA to comply with the requirements of the Council Environmental Quality Regulations for Implementing the Procedural Provisions of NEPA. The FAA reviews licensing applications with an unwavering emphasis on safety, particularly public safety. In addition to scrutinizing a prospective licensee's safety procedures, it examines the proposed flight and accident investigation plans.

Once granted, a license may be suspended or revoked to protect public health and safety, the safety of property, or national security or foreign policy interests.[39] Also, a license may be suspended if a previous launch or reentry under the license has resulted in serious or fatal injury and the Secretary of Transportation determines that continued operations under the license are likely to cause additional serious or fatal injury to the crew or passengers.[40]

Experimental Permits for Test Flights of Reusable Suborbital Spacecraft Not Carrying Passengers or Payloads For Compensation

In establishing a more limited federal regulatory structure for certain reusable suborbital space vehicles, the 2004 Act provides for an experimental permit system designed to streamline the approval process for these craft.[41] Launch operators under the experimentation permit system cannot, however, carry any property or humans for compensation.[42] Conversely, once a license is issued for a specific reusable suborbital rocket design, it cannot be operated under an experimental permit.

An experimental permit allows an unlimited number of launches and reentries by a specific reusable suborbital vehicle design within one year after the issuance of a permit (with yearly renewals) solely for: (1) research and development to test new designs, equipment, or operating techniques; (2) demonstrating compliance with the requirements for obtaining a license; or (3) crew training prior to obtaining a license.[43] The permit system thus eliminates the burden and cost of securing a new license for each test flight.

The act defines a "suborbital rocket" as "a vehicle, rocket-propelled in whole or in part, intended for flight on a suborbital trajectory, and the trust of which is greater than its lift for the majority or the rocket-powered portion of its ascent."[44] In other words, this type of vehicle must fly vertically more than it flies horizontally when its rocket is turned on. The definition of suborbital rocket is not necessarily a permanent one, however. The act provided that in December 2007 (and thereafter) the Secretary of Transportation may issue final regulations changing the definition of a suborbital rocket.[45]

In addition to protecting public health and safety, the safety of property, and national security and foreign policy interests, the act authorizes the establishment of procedures for safety approvals for spacecraft under the experimental launch permit system.[46] Heeding the desire of Congress for a more streamlined approach, with permits granted more quickly and with fewer requirements than licenses, under FAA guidelines, an applicant for an experimental permit must submit: a program description; a flight test plan; and operational safety documentation. An applicant must, however, provide the FAA with sufficient information to analyze the environmental impacts associated with the proposed reusable suborbital rocket flights.[47]

Regulatory Requirements Common To Licenses and Permits

The 2004 act distinguishes crew[48] from passengers, called space flight participants.[49] The act mandates training and medical requirements with respect to the crew, who must receive training and pass medical standards determined by the FAA and must be fully informed in writing that the United States government has not certified the launch vehicle as safe.[50]

Although the crew need not waive claims against their employer and other private entities, such as contractors and subcontractors, passengers fly at their own risk and knowingly assume the risk of spaceflight hazards. The holder of a license to carry passengers into space for compensation must inform them in writing about the "risks of the launch and reentry, including the safety record of the launch or reentry vehicle type. . ." as well as the fact that the United States government has not certified the launch vehicle as safe for carrying crew or space flight participants.[51] After being so informed, passengers must give their written consent to participate in the activity.[52] Prior to December 2007, the FAA had the power to issue regulations requiring physical examinations for passengers; thereafter, the FAA has the authority to set reasonable requirements to protect passengers.[53]

Through December 2012, the 2004 act provides that with respect to launches carrying passengers for compensation the FAA can only restrict or prohibit design features or operating practices that: (1) have resulted in a serious or fatal injury or (2) could have contributed to an unplanned event (or series of events) that pose a high risk of causing serious or fatal injury to the crew or licensed or permitted commercial human space-flight participants.[54] The eight-year window is intended to permit safety standards to evolve in the space flight industry and allow for the revision of these standards. Because one accident or fatality would significantly harm the entire private sector spaceflight endeavor, the industry has a strong incentive to self-regulate and exercise care with respect to safety. During this period, the FAA may only regulate the design or operation of a launch vehicle to protect the public health and safety, the safety of property, or national security and foreign policy interests.[55] Beginning in December 2012, full control of the U.S. commercial spaceflight industry, including the protection of the health and safety of crew and passengers, will pass to the FAA.[56]

Liability Considerations

Zero risk in space activity represents an unattainable goal. The recurrence of accidents with the U.S. Space Shuttles Challenger and Columbia shows that spaceflight remains a risky endeavor, even after detailed investigations and extension reengineering. The review carried out by the Columbia Accident Investigation Board concluded that the "operation of the Space Shuttle, and all human spaceflight is a development activity with high inherent risks.[57] Furthermore, the Findings section of the

2004 act specifically states: "space transportation is inherently risky. . . ."[58] To the extent that private sector space projects find public support, the public must accept risk, specifically, the possibility that lives will be lost and property will be damaged. In this context, both transnational and U.S. laws contain liability regimes.

Transnational Liability Standards

The 1972 Liability Convention addresses the general principles set forth in the Outer Space Treaty regarding a nation's responsibility and liability for damage caused by its space objects. Generally speaking, liability is imposed on a signatory nation from whose territory or facility a space object is launched. A space object may be owned by the public or private sector. The liability runs to other parties to the 1967 treaty, individuals or entities of such nations, for damages caused by a space object (or its component parts) on the Earth, in air space, or in outer space.[59]

The Liability Convention provides specific rules for both personal injury and property damage resulting from space exploration and settlement and for the resolution of those issues on the transnational level. Different liability standards exist depending on where the damage occurs.

The Liability Convention states that a nation, which launches or procures the launching of a space object, or from whose territory a space object is launched, is absolutely liable for damages caused by this space object on the surface of the Earth or to aircraft in flight, without any proof of fault.[60] Under this provision, a victim, provided he, she, or it is not grossly negligent or acting (or omitting to act) intentionally to cause the damage,[61] need only prove a relationship between the action (or inaction) of the defendant and the damage suffered. The defendant cannot raise any defenses.

With respect to damages elsewhere than on the surface of the Earth (or to aircraft in flight) to a space object of one launching nation (or to the persons or property on board such space object) caused by a space object of another launching nation, the launching nation is liable on the basis of fault.[62] This type of liability requires a claimant to prove: the defendant owed him, her, or it a duty of care; the defendant breached that duty of care; and a relationship existed between the breach and the damage caused.

For purposes of either the absolute or fault liability system, the amount of compensation to be paid is vaguely stated as determined under international law and general principles of justice and equity to restore the injured party to the condition as if the damage had not occurred.[63] Also, following the 1967 treaty, the definition of a "launching State" in the Liability Convention considers the nature of a launch vehicle's ownership, whether public or private, irrelevant.[64]

Note that liability attaches to the launching nation. If a space object is sold, so that it no longer remains under the control of the original launching nation, presumably contracts of sale will deal with the liability issue on an ad hoc basis.

With respect to personal injury or property damage, the Liability Convention provides elaborate procedures for asserting claims. In brief, to bring claims on behalf of its business entities or individuals, a nation must present these claims to the launching jurisdiction through diplomatic channels within one year from the date on which the alleged damage occurred. If the parties fail to reach a settlement within one year from the date on which the launching nation receives the claim, then the parties must establish a three-member claims commission which must decide the merits of the case, awarding compensation by a majority vote within one year from the date of its establishment.[65]

With respect to issues apart from personal injury or property damage, such as wrongful death claims, the Outer Space Treaty and the Liability Convention lack a specific dispute resolution mechanism. Assuming both the launching nation and the nation asserting a claim are members of the United Nations, the 1967 treaty indicates that the signatory parties must carry on activities "in the exploration and use of outer space, including the Moon and other celestial bodies, in accordance with international law, including the Charter of the United Nations. . . ."[66] The U.N. Charter states that the parties shall first "seek a solution by negotiation, enquiry, mediation, conciliation, arbitration, judicial settlement, resort to regional agencies or arrangements, or other peaceful means of their own choice."[67] If these means fail to resolve the issue, the charter further provides that "legal disputes should as a general rule be referred by the parties to the International Court of Justice. . ." for a decision in that forum.[68] If a dispute cannot be resolved by either of these methods and the dispute endangers international peace and security, then the charter requires the parties to refer the matter to the U.N. Security Council.[69]

The United States declared its acceptance of the International Court of Justice's jurisdiction under the U.N. Charter in August 1946.[70] However, the United States withdrew its acceptance in 1985 in response to the World Court's decision in *Nicaragua v. United States*.[71] Therefore, the United

States, among other nations, cannot bring certain space disputes to the International Court and in the absence of an agreement creating binding procedures for these grievances, these nations will most likely seek to resolve these controversies through diplomacy. However, with respect to claims against the United States as a launching nation, other nations, private entities, and individuals may file claims in American courts or administrative agencies.[72]

From this brief summary, one point becomes apparent. A need exists for an easier system for the adjudication of claims between (and among) parties operating in space. The Liability Convention ought to be revised to provide a workable dispute resolution mechanism. American businesses and individuals do not want to rely on the United States government to assert liability claims on their behalf.

Because liability under the Liability Convention rests on the launching nation for a claim resulting from a private party's action, it necessitates a licensing and regulatory system under national laws and regulations to protect governmental funds. Thus, a number of space-faring nations, including the United States, have enacted laws dealing with the financial responsibilities of their private sector space enterprises.

United States Laws Regarding the Liability of Licensees and Permittees

Under American law, a licensee or a permittee,[73] must obtain liability insurance or demonstrate financial responsibility to compensate for the maximum probable loss from claims related to a launch or reentry to third parties for death, bodily injury or property damage or to the United States government, but not to passengers or crew who make their own risk decisions. The FAA makes the maximum probable loss determination, that is, a launch vehicle's most probable loss for damage, injury, or loss.[74] To meet this requirement, a licensee or a permittee will typically obtain liability insurance to compensate for claims by a third party—but not a passenger or crew member, for death, bodily injury, or property damage or loss resulting from an activity carried out under the license or a permit—in the amount of $500 million for the total claims related to one launch or reentry.[75] A licensee or a permittee will also typically obtain liability insurance to compensate for claims brought by the United States government for damage or loss to federal government property resulting from an activity carried out under a license or permit in the amount of $100 million for the total claims related to one launch or reentry.[76] In any event, the maximum amount of insurance coverage is subject to the maximum available on the world market at a reasonable cost.[77]

To lessen the liability exposure for spaceflight operations and to coverage catastrophic claims, the United States government, subject to congressional appropriation, provides indemnification against a successful claim by a third party, but not a passenger or crew member, resulting from an activity carried out under a license (but not a permit) for death, bodily injury, or property damage or loss up to a maximum of $1.5 billion (as indexed for inflation) above the amount of the required insurance coverage.[78] Indemnification by the federal government to commercial space flight licensees for liability to third parties helps support the nascent industry by protecting against high insurance costs due to the risk of a single catastrophe. Indemnification for claims above the insurance coverage or financial responsibility amounts applies to licenses issued for applications filed on or before December 31, 2009.[79] Financial responsibility above the combined amount of insurance and indemnification is imposed on a licensee unless it can show an absence of liability.

Pursuant to a license issued by the FAA, parties must execute reciprocal waivers of claims, thereby requiring private sector space enterprise participants—including a licensee, its customers, and contractors, subcontractors of a licensee and its customers—to negotiate their own liability arrangements.[80] Thus, the risk management process for private sector spaceflight will involve the allocation of risk through: reciprocal waivers of liability among these participants (except for a party's willful misconduct); commitments to obtain insurance; indemnification granted by the United States; and various negotiated limitations on and exclusions from liability.

Property Rights and Resource Appropriation Under Transnational Law

With the prospect of mining activities on the Moon and various asteroids as well as efforts to tap solar power and transmit it back to the Earth and various settlements on the Moon, the Outer Space Treaty seeks to ensure free access to outer space by preventing the clash of sovereign property claims to space and its resources.[81] Article I of the Treaty declares that the exploration and use of outer space "shall be carried out for the benefit and in the interests of all countries, irrespective of their degree of economic or scientific development, and shall be the province of all mankind."[82] Furthermore, the exploration and use of outer space is to be free from any discrimination, with free access accorded to all areas of celestial bodies. By prohibiting national territorial sovereignty over any part of outer space, the treaty provides for an international sovereignty regime. Article II specifically states: "Outer space, including the moon and other celestial bodies, is not subject to national appropriation by claim of sovereignty, by means of use or occupation, or by any other means."[83] Thus, no government can claim sovereignty over any part of outer space or any celestial body.

Because of the language prohibiting national sovereignty and ownership, but the notable lack of such language prohibiting private ownership, some commentators maintain that business entities and individuals may claim ownership of all or part of celestial bodies.[84] One provision of the Outer Space Treaty states that it applies to joint activities by nations that are party to the treaty, whether singly or jointly, and any international inter-governmental organizations.[85] If the drafters of the treaty made the point of writing an extra provision to cover this possibility they could have also drafted a specific prohibition of property ownership by private entities. No such specific language exists in the Outer Space Treaty, lending support to the idea that while nations may not stake claims to territory in outer space, private companies and individuals may do so.

However, other commentators conclude that the inclusion of the phrase "by any other means" precludes both business entities and individuals from making property claims in outer space.[86] Use of the phrase "all countries" regardless of "their degree of economic or scientific development" may indicate that the ownership of the resources in outer space, even sunlight, is jointly held by entire planet and thus the benefits must be shared by all nations. Furthermore, as considered earlier in this chapter, anything launched into outer space is deemed the responsibility of the launching nation for liability purposes. Because this provision applies to any payload launched or settlement made in outer space by a private firm, arguably a private company could not hold property rights there. Under the Outer Space Treaty, a business entity is considered an arm of the nation from which it was launched.

Ironically, U.S. firms may be out of luck even if the Outer Space Treaty permits property rights for private entities. Under American law, which is based on English common law, the United States must have sovereignty over territory before being able to confer a title to real property on private entities. Because the Outer Space Treaty specifically prohibits national territorial sovereignty and under American common law principles property rights derive from territorial sovereignty, the United States seemingly could not confer property rights on its private entities.

However, the Outer Space Treaty does not rule out all possible private property claims in outer space. Three possibilities exist on which to base private property rights without national sovereignty. First, property rights may derive from the jurisdictional peg conferred by the treaty, which provides that property rights to objects launched into space do not dissolve once they have obtained orbit, thereby establishing a quasi-territorial jurisdiction. When a firm launches tangible property into space, it remains the entity's private property. As the treaty states: "Ownership of objects launched into outer space, including objects landed or constructed on a celestial body, and of their component parts, is not affected by their presence in outer space or on a celestial body or by their return to the Earth."[87] Thus, the United States could confer ownership rights on structures—whether on the Moon or asteroids or in space, built by private firms, whether from earthly or locally available materials—even in the absence of territorial sovereignty. This line of reasoning has significance for Bigelow's orbiting, inflatable habitats, discussed in Chapter 6. Bigelow's habitats comprise his firm's private property when launched and would remain his entity's private property even when they reach orbit.

Second, under the treaty, nations likely have jurisdiction over space facilities constructed by their business entities on the Moon, the asteroids or in space. Thus, a nation may enact laws governing the mining, research, or manufacturing activities of its business entities and citizens as well as these activities within space facilities and safety zones on their registry.[88] Furthermore, a nation could exercise its jurisdiction over areas under the control of its business entities and citizens to prevent interference by others and to ensure safety, together with a reasonable area around such facilities and to personnel in or near such facilities, irrespective of their nationality. Because a nation would have the right to control the activities of all persons and legal entities within an owner's space facility and safety zone, it could exclude others from these facilities and safety zones.

However, the extent of these zones would be limited in order to maintain free access to space to other nations which are party to the Outer Space Treaty. A nation's jurisdictional control would also be limited in time. Its jurisdiction with respect to a given area would terminate when a facility—for example, a research laboratory—returns to Earth, is destroyed or abandoned, or when activity stops inside or outside of it.

Third, although it may be argued that resources on the Moon or an asteroid are part of the land surface and cannot be treated separately from it, thereby barring private entities from claiming ownership of mineral resources "in place," once mined from a space facility and its accompanying safety zone, these resources seemingly become subject to private ownership. The appropriation of natural resources may follow from the Outer Space Treaty's provision for the freedom of exploration and use in outer space.[89] However, the same article also mandates that exploration and use are permissible only to the extent they are "for the benefit and in the interests of all countries. . . ."[90]

Although undoubtedly the worst case scenario is one where commercial entities would be denied property rights, the current situation, an uncertain one where the Outer Space Treaty could go either way with respect to private property rights, has a significant and detrimental effect on the market. The ambiguities of the 1967 treaty that remain forty years after its entering into force may plague the private sector, particularly if Third World nations, worried about a new form of imperialism, press to define space resources as common property.

With respect to resource extraction and utilization, the Moon Treaty must also be considered. In an attempt to further delineate space activities and to provide a measure of equity for less economically developed nations, the Moon Treaty declares: "The moon and its natural resources are the common heritage of mankind. . . ."[91] This treaty not only applies to the Moon, but also indicates that its provisions apply to other celestial bodies in the solar system, except the Earth.[92]

The United States is not a signatory to the Moon Treaty, which entered into force with respect to the ratifying nations on July 11, 1984. In fact, no space-faring nation is a signatory to this treaty because of its extremely liberal redistribution policy. Instead, an odd collection of thirteen nations have ratified the Treaty and four other have signed, but not ratified, it, including France and India.[93] Although the space-faring nations signed the Outer Space Treaty, with its overtones of peaceful use and vague property rights, the Moon Treaty represents a completely different commitment, one requiring signatory nations to give up any advantage they may gain for the benefit of all. Thus, the few nations that had developed the technology to get to the Moon and beyond, such as the United States and the then Soviet Union, refused to sign or ratify this treaty. The treaty's provisions were too extreme even for the former U.S.S.R.

Mandating the joint ownership of space resources and a duty to share profits, the Moon Treaty states that the "exploration and use of the moon shall be the province of all mankind and shall be carried out for the benefit and in the interests of all countries, irrespective of their degree of economic or scientific development" and that consideration must be paid to the interests of present and future generations and to the promotion of "higher standards of living and conditions of economic and social progress and development. . . ."[94] Similar to the Outer Space Treaty, the Moon Treaty provides that, "the moon is not subject to national appropriation by any claim of sovereignty, by means of use or occupation, or by any other means."[95] However, the Moon Treaty removes any ambiguity as to lunar property rights, stating:

> Neither the surface nor the subsurface of the moon, nor any part thereof or natural resources in place, shall become the property of any State, international intergovernment or non-governmental organization, national organization or non-governmental entity or any natural person. The placement of personnel, space vehicles, equipment, facilities, stations and installations on or below the surface of the moon, including structures connected with its surface or subsurface, shall not create a right of ownership over the surface or the subsurface of the moon or any areas thereof."[96]

Following this philosophy, the treaty provides for the establishment of an international regime to "govern the exploitation of the natural resources of the moon. . . ."[97]

In sum, the Moon Treaty prohibits any private property rights on the Moon and contemplates the development and removal of the Moon's natural resources under the management of an international regime established for that purpose, which would decide what resources to pursue and how to distribute the resources and the revenues from these resources. By precluding private rights and profits, it negates the impetus for commercial development of the Moon. Simply put, the Moon Treaty is unacceptable to space-faring nations in light of the risks involved in getting to the Moon and extracting its resources. The treaty would prevent the development of lunar resources that could help all humanity. Along with the suggested revisions to the Outer Space Treaty, the Moon Treaty ought to be abrogated.

In coming years, policymakers and legislators must regularly revisit the existing U.S. regulatory regime to see if it limits American private sector entry into the space market by imposing burdensome and costly licensing, among other, requirements. The public must accept that space exploration is a risky endeavor and that lives will be lost. The federal government must continue to provide indemnification coverage against a successful claim by a third party resulting from an activity carried out under a space license. The United States ought to continue to modify and simplify its permit and licensing system and not impose restrictive liability provisions.

Turning to transnational regulation, although much about the Moon and various asteroids is still unknown, multitudes of possible resources may exist to be extracted and numerous possible ways may exist for the private sector to generate profits. These resources include solar power generating satellites, lunar mining for elements such as aluminum, silicon, oxygen, and a rare and highly valuable isotope of helium, and asteroid mining for rare and precious metals. Because so little is known about what can be harvested from the Moon and other celestial bodies, private sector investments will be risky. This is made even more so by the Outer Space Treaty's unpredictable legal terrain, most notably whether private property rights exist in outer space and whether this treaty applies only to nations and public sector entities. Simply put, the Outer Space Treaty fails to provide a positive regime for private sector space development.

While much of current space commerce revolves around satellites and orbital slots, future celestial investment may center on resource extraction. For this reason, the Outer Space Treaty itself (or through ancillary treaties) ought to be updated to provide property rights for commercial entities. The present vagueness with respect to property rights, an inability to claim territory, and the possibility that profits must be shared with all nations, adds to the financial risk of space exploration, extraction, and settlement activities. The report of the U.S. Presidential Aldridge Commission concluded that the uncertainty created by disallowing property rights in space "could strangle a nascent space-industry in its cradle; no company will invest millions of dollars in developing a product to which their claim is uncertain."[98]

A degree of realism is in order, however, with respect to the possibility of amending the Outer Space Treaty. Desirous of sharing the benefits of resource appropriation, Third World nations will likely fight changes designed to promote private sector space activities. However, space resources may be nearly limitless, thereby negating the need for benefit sharing, such as done in the Moon Treaty. Even if developing nations take decades to be able to tap space resources, these resources will likely remain plentiful. Also, as noted in Chapter 1, activities in outer space provide a number of unique opportunities, including unlimited solar energy as well as a microgravity environment and a near vacuum for scientific research and various manufacturing processes. Thus, if the resources pan out and private firms are able to turn a profit, the transnational law surrounding mining will likely be redefined, to provide commercial entities with property rights, as it will apply to much more than a handful of moon rocks that came home on the Apollo 11.

Even if the Outer Space Treaty is not modified, but the private sector races ahead of transnational legal changes, an international registry for property rights and mineral claims may come into existence. The special transnational body, not part of the United Nations, but modeled after other global entities, such as the World Trade Organization, would serve two purposes. It would not only record real property rights and mineral claims but also allow commercial entities to demonstrate a feasible plan for the extraction of outer space resources, recognizing the likely role of remote sensing and robotic probes, at least initially. This type of registry would assure private developers and settlers that other entities, whether from the public or private sector, would respect their rights. Presumably private claims would be limited to areas actually used, thus not removing large tracts from future development.

Notes

1. Department of State, United States of America, *United States Treaties and Other International Agreements*, Volume 18, Part 3 (Washington, DC: U.S. Government Printing Office, 1969), 2410-2421, entered into force on October 10, 1967.
2. *United States Treaties*, Volume 19, Part 6 (1969) 7570-7577, entered into force on December 3, 1968.
3. *United States Treaties*, Volume 24, Part 2 (1974), 2389-2404, 2389, entered into force on September 1, 1972.
4. *United States Treaties*, Volume 28, Part 1 (1978), 695-703, entered into force on September 15, 1976.
5. United Nations General Assembly Resolution 34/68, A/34/68 (December 5, 1979), entered into force on July 11, 1984.
6. United Nations treaties and principles on outer space related General Assembly resolutions, Status of international agreements relating to activities in outer space as of 1 January 2007, March 2007, ST/SPACE/11/Rev.1/Add.1/Rev. 1.
7. Outer Space Treaty, Preamble.
8. *Ibid.*, Article I.
9. For an English translation of Bogota Declaration see *Journal of Space Law* 6:2 (Fall 1978): 193-196.
10. Outer Space Treaty, Article IV.
11. *Ibid.*, Article V.
12. *Ibid.*
13. Rescue Agreement, Articles 1-4. The term "launching authority" is defined in *Ibid.*, Article 6.
14. Outer Space Treaty, Article VI.
15. *Ibid.*, Article VIII.

16.   Registration Convention, Article II, paragraph 1. The definition of the term "space object" contained in *Ibid.*, Article I(b) likely does not include an orbiting habitat.

17.   Public Law 99-514, Section 1213(a), codified at 26 USC §863(d).

18.   Public Law 101-580, Section 1(a), codified 35 USC §105.

19.   Public Law 97-96, Section 6, codified at 18 USC §7(6).

20.   Public Law 98-375.

21.   National Security Decision Directive 42 (July 4, 1982), "National Space Policy," in Christopher Simpson, *National Security Directives of the Reagan and Bush Administrations: The Declassified History of U.S. Political and Military Policy,* 1981-1991 (Boulder, CO: Westview, 1995), 136-143 (classified version), 144-150 (declassified version).

22.   *Ibid.*, 136, 144.

23.   *Ibid.*, 137, 145.

24.   *Ibid.*, 140, 147. The Reagan administration also issued NSDD 144, "National Space Strategy," on August 15, 1984, calling on the federal government to "encourage the private sector to undertake commercial space ventures without direct Federal subsidies" by eliminating or revising "discriminatory" tax laws and regulations and updating laws and regulations "to accommodate space commercialization." Simpson, *National Security Directives,* 419.

25.   Executive Order 12465, "Coordination and Encouragement of Commercial Expendable Launch Vehicle Activities," *Federal Register* 49:40 (February 24, 1984): 7211-7212.

26.   Public Law 98-575, Sections 4(2)-(6) and 6(a)(1).

27.   *Ibid.*, Section 9(a) and (b).

28.   *Ibid.*, Section 3(1).

29.   *Ibid.*, Section 3(2).

30.   Public Law 100-657, Section 5.

31.   Public Law 105-303, Section 102.

32.   Public Law 108-492.

33.   Codified at 49 USC §70101(a)(7).

34.   *Congressional Record* 150:134-Part II (November 20, 2004): 10097 and *Ibid.* 150:139 (December 8, 2004):S12029.

35.   49 USC §70101(a)(13).

36.   49 USC §§70104(a) and 70102(1).

37.   Federal Aviation Administration regulations provide for two types of licenses, RLV launch-specific licenses and RLV site operator licenses. FAA Commercial Space Transportation Regulations, *Code of Federal Regulations,* Volume 14, Part 415.3. The FAA review process is set forth in *Code of Federal Regulations,* Volume 14, Part 415, specifically, Parts 415.21-415.103.

38.   42 USC §4321 et seq.

39.   49 USC §70107(c)(2). See also 49 USC §70108(a).

40.   49 USC §70107(d)(1). The terms "serious or fatal injury" is as defined in *Code of Federal Regulations,* Volume 49, Part 830.2, as in effect on November 10, 2004.

41.   49 USC §70105a.

42.   49 USC §70105a(h).

43.   49 USC §70105a(d) and (e).

44.   49 USC §70102(19). The term "suborbital trajectory" is defined in 49 USC §70102(20); "launch" in 49 USC §70102(4); "launch vehicle" in 49 USC §70102(8); "reentry" in USC §70102(13); "reentry vehicle" in 49 USC §70102(16).

45. 49 USC §70120(c)(2)(A).
46. 49 USC §70105a(b).
47. Federal Aviation Administration, Experimental Permits for Reusable Suborbital Rockets, Notice of Proposed Rulemaking, *Federal Register* 71:62 (March 31, 2006): 16251-16273.
48. The term "crew" is defined in 49 USC §70102(2).
49. The term "spaceflight participant" is defined in 49 USC §70102(17).
50. 49 USC §70105(b)(4).
51. 49 USC §70105(b)(5)(A) and (B).
52. 49 USC §70105(b)(5)(c).
53. 49 USC §70105(b)(6).
54. 49 USC §70105(c)(2).
55. 49 USC §70105(c)(4). FAA regulations contain certain design requirements, specifically, environmental control and life support systems as well as smoke detection and fire suppression. *Code of Federal Regulations*, Volume 14, Parts 460.11 and 406.13.
56. 49 USC § 70105(c)(3).
57. Columbia Accident Investigation Board, *1 Columbia Accident Investigation Board Report,* Report Volume I (Washington, DC: National Aeronautics and Space Administration and U.S. Government Printing Office, 2003), 9.
58. 49 USC §70101(a)(12).
59. Outer Space Treaty, Articles VI and VII.
60. Liability Convention, Articles I and II. The term "space object" is defined in *Ibid.*, Article I(d).
61. *Ibid.*, Article VI, Paragraph 1.
62. *Ibid.*, Article III.
63. *Ibid.*, Article XII.
64. *Ibid.*, Article I(c).
65. *Ibid.*, Articles VIII, IX, X, XIV, XV, XVIII.
66. Outer Space Treaty, Article III.
67. Charter of the United Nations, Article 33.
68. *Ibid.*, Article 36(3).
69. *Ibid.*, Article 37(1).
70. Declaration On The Part Of The United States of America, August 14, 1946.
71. Statement, U.S. Terminates Acceptance of ICJ Compulsory Jurisdiction, October 7, 1985.
72. Liability Convention, Article IX, Paragraph 1.
73. 49 USC §70105a(I).
74. 49 USC §70112(a)(2) and (c).
75. 49 USC §70112(a)(3)(A)(I). The term "third party" is defined in 49 USC §70102(21).
76. 49 USC §70112(a)(3)(A)(ii).
77. 49 USC §70112(a)(3)(B).
78. 49 USC §70113(a).
79. 49 USC §70113(f).
80. 49 USC §70112(b).
81. I have drawn on Wayne N. White Jr., "Real Property Rights in Outer Space," Proceedings, 40th Colloquium on the Law of Outer Space, American Institute of Aeronautics and Astronautics (1998) <http://www.spacefuture.com/archive/real_property_rights _in_outer _space.shtml> (June 19, 2006). See also Wayne White, "The Legal Regime for Private Activities in Outer Space," in *Space: The*

Free-Market Frontier, ed. Edward L. Hudgins (Washington, DC: CATO Institute, 2002).

82.   Outer Space Treaty, Article I.
83.   *Ibid.*, Article II.
84.   See, e.g., Stephen Gorove, "Interpreting Article II Of The Outer Space Treaty," *Fordham Law Review* 37:3 (March 1969):349-354, at 351-352; Stephen Gorove, *Developments in Space Law* (Dordrecht, The Netherlands: Martinus Nijhoff Publishers, 1991), 114-115; Glenn Reynolds, "Space Law in the 1990s: An Agenda for Research," *Jurimetrics* 31:1 (Fall 1990): 1-15, at 5.
85.   Outer Space Treaty, Article XIII, Paragraph 1.
86.   See, e.g., Lawrence A. Cooper, "Encouraging Space Exploration through a New Application of Space Property Rights," *Space Policy* 19:2 (May 2003): 111-118, at 112.
87.   Outer Space Treaty, Article VIII.
88.   *Ibid.*
89.   *Ibid.*, Article I, Paragraph 2.
90.   *Ibid.*, Article I, Paragraph 1. See Bin Cheng, *Studies in International Space Law* (Oxford: Clarendon Press, 1997), 233-234.
91.   Moon Treaty, Article 11, Paragraph 1.
92.   *Ibid.*, Article 1, Paragraph 1.
93.   United Nations treaties and principles on outer space and other related General Assembly resolutions, Status of international agreements relating to activities in outer space as of 1 January 2007.
94.   Moon Treaty, Article 4, Paragraph 1.
95.   *Ibid.*, Article 11, Paragraph 2.
96.   *Ibid.*, Article 11, Paragraph 3.
97.   *Ibid.*, Article 11, Paragraph 5.
98.   Report of the President's Commission on Implementation of United States Space Exploration Policy, *A Journey to Inspire, Innovate, and Discover,* June 2004, 34.

8

Conclusion

The market for human spaceflight seems ripe for commercial development. Innovative firms, featured in Chapters 3 through 6, have found start-up financing. Assuming that one or more of these companies achieve technical success, demonstrate financial viability, and avoid catastrophic accidents, they will likely obtain significant amounts of private investment funds, from venture capital firms and the public offerings of their securities, among other sources.

Space exploration—beginning with suborbital space tourism—offers a promising new area of commercial activity. Given the opportunity, people will travel into space, with the demand for tourism likely serving as the catalyst for a commercial space industry as the cost to access space decreases. With a sound national and transnational regulatory environment, entrepreneurs would generate significant profits. However, a need exists to bring down launch costs to levels that actualize the multi-billion dollar commercial potential. Hopefully, the fledgling private sector companies will drive down the high cost of launching people and payloads into space and continue to develop innovative technologies. Advances in space exploration will likely result in a significant drop in launch costs and greater vehicle reliability.

Beyond human spaceflight, if launch costs decrease from $10,000 per pound, then to $1,000 per pound and as low as $100 per pound, numerous other commercial possibilities become economically viable at orbiting habitats, including various forms of manufacturing and product-testing, facilities for high-density, high-intensity agricultural production, and orbiting health and medical facilities, useful for heart disease and burn victims, whose recovery would benefit from a microgravity environment. Businesses may also demonstrate that resources on the Moon or asteroids are valuable enough to use them on-site, in orbit, or return them to Earth.

Further research and development will be needed with respect to mining and transporting these resources.

To tap this potential, launch vehicles need to undergo a period of evolution. By starting with smaller, reusable vehicles, entrepreneurs can use the lessons learned from their operation to then build larger vehicles.

However, going from suborbital flight to getting people to and from orbital space, cheaply and safely, represents an enormous leap. According to Elon Musk, "It's like building something to cross the English Channel and one to cross the Atlantic."[1] Because the speed needed to get into orbit is some eight times the velocity needed to reach sixty-two miles from Earth, the propulsive energy required for orbital flights is some sixty-four times of what is needed for suborbital flights. Thus, the need to go from zero to 17,000 mph to get out of the Earth's gravity and go into orbit requires a powerful rocket engine.

Generally speaking, a suborbital reusable launch vehicle (RLV) would not endure the same reentry hazards as an orbital RLV. The energy needed to be dissipated as heat on reentry of an orbital vehicle is significantly greater with fifty times more energy dissipated on the reentry of an orbital RLV than on a suborbital RLV. However, orbital passenger vehicles will experience lower thermal stress on reentry than multi-purpose spacecraft. Passenger vehicles will start to decelerate higher in the upper atmosphere, thereby experiencing lower peak reentry heating loads. In contrast, the Space Shuttle, for example, decelerates lower in the atmosphere, thereby leading to severe peak heating.

Even with lighter spacecrafts and more efficient rocket engines, venturing beyond the Earth's orbit is exponentially more costly and dangerous. Rather than looking to rocket engine propulsion, which is space consuming, requiring huge amounts of fuel, a technological breakthrough may be required.

One possibility is a space elevator (called a "lifter"), a very long, lightweight cable or ribbon, made out of strong carbon nanofibers, that starts on the Earth and rises for some 62,000 miles (100,000 kilometers) up into space.[2] The elevator would extend from an offshore, sea platform on the Pacific Ocean near the Equator up to an orbiting satellite. The cargo, whether human, robots, or payload, would crawl along a vehicle into space without a rocket-propelled spacecraft. Laser beams striking solar cells on board could power the space elevator; with a receiver obtaining additional power from a big laser on the ground and an optical system to transmit the beam. Doing away with on-board rocket fuel would significantly reduce the cost of moving humans or cargo into outer space.

The reusable elevator, perhaps the size of a railroad car, would shuttle up and down the ribbon, with mechanical rollers gripping the long, thin cable ferrying people and cargo up into space at a steady 120 miles (200 kilometers) per hour. It would serve as a high capacity vehicle capable of sending thousands of tons of supplies every year into orbit and then on to the Moon.

Nanotechnology provides the technological key to the space elevator. The ribbon would be made of light, but superstrong, fibers woven from durable carbon nanotubes, built from tiny, molecular threads of carbon atoms, one ten-thousandth the diameter of a human hair in circumference and one-thousandth a hair's diameter in length. The ribbon of nanotubes, as envisioned, would be three feet wide and thinner than a sheet of paper.

The ribbon would ultimately hang from a man-made counterweight, an anchor, orbiting 62,000 miles (100,000 kilometers) up in space, which would hold the entire system in place. The ribbon would connect to an aircraft-carrier sized platform floating in the Pacific Ocean. Initially, another platform would be built 22,300 miles up, the height of the Earth's geosynchronous orbit, where satellites align with the earth's rotational speed to circle the planet once a day. The space elevator would rely on the gravity and the centrifugal force of the Earth's rotation to hold the long, thin ribbon taut.

Getting the ribbon "up there" is called deployment. A rocket would launch the first (seed) ribbon to the requisite altitude, which then would spool down, followed by more ribbon launches.

Beyond the technological hurdles, financial, legal, and security obstacles exist for the implementation of the space elevator. The first is funding for the project. Although the entire system could be built and deployed for an estimated $10-$12 billion,[3] private investors seem leery of financing a highly speculative project that will take decades for implementation. Firms interested in building a space elevator will need to commercialize spin-offs from the development as the key to finance their efforts. Apart from any public sector involvement, profits from these spin-offs will help fund the further development and commercialization of the space elevator.

The political position of equatorial nations' 1976 Declaration of Bogota, considered in Chapter 7, may impact on the plans for the space elevator. It is unclear whether these countries will seek to claim a piece of the pie.

The space elevator could be a potential target for strikes by terrorist groups or rogue nations. Appropriate design and operational measures need to be taken to make any attack as difficult as possible, thereby reducing the chances that the space elevator would become a target. Anchoring the ribbon for the space elevator in a remote region of the Pacific Ocean would enable its perimeter to be more effectively guarded.

A danger also exists from orbiting space debris and natural micrometeoroids that could damage or destroy the cable. Radar could detect the debris and other objects before a collision and enable the anchor ship to move to avoid the incoming pieces. The ribbon itself would be resilient to small space debris damage.

Despite the technological, financial, legal, and security obstacles, at least three American companies, LiftPort, Inc., Carbon Designs, Inc., and X-Tech Projects, beyond the scope of this book, exist and are pursuing commercial space elevator projects. NASA has studied the concept, which is coming within practical reach as a result of advances in carbon nanotube manufacturing, and is sponsoring the development of space elevator-related technology as part of its Centennial Challenges prize program, discussed in Chapter 2. NASA has funded the Climber (Power Beaming) and the Tether Strength competitions, part of the annual Space Elevator Games, to encourage research and development in this area.[4]

Is the space elevator the stuff of science fiction? Yes. It was made famous by Arthur C. Clarke in his 1979 novel, *The Fountain of Paradise*.[5] Once deployed, however, the space elevator will provide low-cost access to space for roughly $100 per pound initially, and substantially less, with increasing volume,[6] thereby paving the way for space exploration, use, and settlement in the twenty-first century.

Investors and business people, such as Allen, Ansari, Bigelow, Branson, and Musk, are willing to put their faith and their funds into space endeavors, with the ultimate goal for many of them of enabling our planet to become a space-faring civilization. In opening the final frontier, these space entrepreneurs assert that humans must eventually explore and settle other planets. According to Elon Musk of SpaceX, "Long-term, our ambition is to help humanity become a multiplanet species."[7] This goal will likely be achieved before the end of this century.

Notes

1.   Mike Tolson, "Private Space Enterprise Lifts Off," *Houston Chronicle*, October 10, 2004, A1.

2.   I have drawn on Lee Gomes, "Is the Final Frontier Just One Ride Away on a Space Elevator?, " *Wall Street Journal*, August 22, 2007, B1; J. Bonasia, "Will An 'Elevator To The Top' Even Mean Truly to the Top?", *Investor's Business Daily*, June 19, 2006, A4; *The Economist*, "Waiting for the Space Elevator," 379:8481 (June 10, 2006): 4; Brad Stone, "Climbing a Ribbon to Space," *Newsweek* 144:17 (October 25, 2004): 64. For a technical explanation of the space elevator, its possibilities and problems, including its social aspects and business impact, see *LiftPort: The Space Elevator Opening Space To Everyone*, eds. Michael J. Laine, Tom Nugent, Bill Fawcett (Decatur, GA: Meisha Merlin Publishing, 2006), 44-48, 61-67, 85-90, 101-107, 119-134, 152-159, 162-222, 265-270. See also Bradley C. Edwards and Eric A. Westling, *The Space Elevator: A Revolutionary Earth-to-Space Transportation System* (Houston, TX: BC Edwards, 2003). The concept of a space elevator, using graphite whiskers, was put forward by an engineering graduate student Yuri Artsutanov, "To the Cosmos by Electric Train," *Young Persons' Pravda*, Sunday Supplement, July 31, 1960, n.p., translated by Joan Barth Urban and Roger G. Gilbertson.

3.   Elevator: 2010, "The Space Elevator Primer" <http://www. elevator2010.org/ primer.html> (December 21, 2006).

4.   National Aeronautics and Space Administration, Press Release, "NASA Announces First Centennial Challenges' Prizes," Release 05-083, March 23, 2005; NASA, Press Release, "NASA Announces Results from Beam and Tether Challenges," Release 05-347, October 24, 2005. See also Elevator 2010's Annual Space Elevator Games <http://www.elevator2010.org/site/competition. html>(December 21, 2006).

5.   Arthur C. Clarke, *The Fountains of Paradise* (New York: Harcourt Brace Jovanovich, 1979). See also Arthur C. Clarke, "The Space Elevator: 'Thought Experiment,' Or Key To The Universe?," Address to the XXXth International Astronautical Congress, September 20, 1979 <http://www.islandone.org/ LEOBiblio/CLARK1. HTM>(November 17, 2006).

6.   Elevator: 2010, "The Space Elevator Primer" <http://www. elevator2010.org/ primer.html> (December 21, 2006).

7.   Kevin Maney, "Private sector enticing public into final frontier," *USA Today*, June 17, 2005, 1B.

Index